Preface

This book is designed to be a supplement for students in a one-term biochemistry course who have not previously taken organic chemistry. If you are a student in such applied biological sciences as agriculture, home economics, nursing, and the paraprofessional health fields, you need an introductory knowledge of biochemistry, but the demands of your specialization may not allow time to take a course in organic chemistry before taking biochemistry. Nevertheless, biochemistry is the chemistry of compounds involved in life processes and these compounds are almost exclusively organic compounds. The language of organic chemistry, therefore, is necessary to discuss biochemistry. *Essentials of Organic Chemistry* supplies the concepts in organic chemistry that are necessary to understand elementary biochemistry.

If used in a course, the material presented here may be covered in the first nine to twelve hours of lecture, leaving the remainder of the term for discussion of biochemistry. A number of elementary biochemistry books are available to serve as texts for the latter portion of the course.

This book is also useful for those students who have had a course in organic chemistry and wish to review the highlights quickly in anticipation of a course in biochemistry or in preparation for standard general examinations. Reading the entire book in one sitting is feasible. *Essentials of Organic Chemistry* will also be helpful to students interested in a brief overview before taking a one- or two-term course in organic chemistry.

The primary emphasis of this book is on the characteristics of organic functional groups, particularly those most common to biological compounds. If used as a preparation for biochemistry it is important that the student master the entire book. The information presented represents the minimum necessary to understand biochemistry.

End-of-chapter exercises, a practice exam, and extra exercises in naming compounds are included to help you determine your mastery of the important concepts. Practical information on reaction sequences and structures of some complex organic compounds is found in the appendix.

The author is grateful to the following reviewers for their careful attention and valuable suggestions: Steven Clarke of The University of California, Los Angeles; Irwin J. Goldstein of The University of Michigan Medical School; Ross Hume Hall of McMaster University; Robert M. Munn of The University of Maryland; David S. Page of Bowdoin College; Henry Z. Sable of Case Western Reserve University; and David B. Shear of The University of Missouri.

A special thanks is due to Mrs. L. Maxine Morton, who has performed the typing from the manuscript's origin as a class handout to its finished form. Being accustomed to her excellent skills for many years does not diminish my admiration for her efforts here.

Lincoln, Nebraska **H.W.K.**

Essentials of Organic Chemistry

Herman W. Knoche
University of Nebraska at Lincoln

ADDISON-WESLEY PUBLISHING COMPANY

Reading, Massachusetts • Menlo Park, California
Don Mills, Ontario • Wokingham, England
Amsterdam • Sydney • Singapore • Tokyo
Mexico City • Bogotá • Santiago • San Juan

This book is dedicated to my wife, Darlene, whose support has been as immutable as any chemical law I know.

Library of Congress Cataloging-in-Publication Data

Knoche, Herman W.
Essentials of organic chemistry.

Includes index.
1. Chemistry, Organic. I. Title.
QD251.2.K66 1985 547 84-28257
ISBN 0-201-11670-7

ABCDEFGHIJ–AL–898765

Contents

Introduction

THE CHEMISTRY OF ORGANIC COMPOUNDS

When the science of chemistry was developing it was believed that only living organisms could make organic compounds. Hence the term *organic* originally meant "from a living organism." It was thought that only the mysterious "vital force," characteristic of life, could convert mineral matter into organic matter. In those days organic chemists were trying to determine the structure of these compounds produced through life. A simple compound, urea, had been isolated from urine, and they knew that it had the empirical formula CH_4NO:

$$\begin{array}{c} O \\ \| \\ H_2N{-}C{-}NH_2 \end{array}$$

urea

But in 1828 the idea of a vital force suffered a serious blow by the demonstration that urea could be made by heating ammonium cyanate (NH_4NCO), which was considered to be a mineral or inorganic compound. Professor Friedrich Wöhler, a German, was the chemist who accomplished this.

Of course we now know that compounds produced by living organisms may be produced by chemical reactions *in vitro* ("in glassware"). So the term *organic compounds* now refers to compounds that contain the element carbon. In fact, some people refer to organic chemistry as carbon chemistry. This is close to being correct, but organic compounds contain other elements, particularly hydrogen, oxygen, nitrogen, phosphorus, and sulfur, and some trace elements such as iron, cobalt, and zinc.

BIOLOGICAL CHEMISTRY OR THE CHEMISTRY OF BIOLOGICAL COMPOUNDS

If the idea of the vital force had not been destroyed, biological chemistry and organic chemistry would mean exactly the same thing. With the redefinition of organic chemistry, much of biological chemistry (or biochemistry) is technically a branch of organic chemistry. Is it necessary to understand the whole field of organic chemistry in order to approach biochemistry? No, but one cannot study reactions that occur in biological tissues and involve organic compounds without having some understanding of those compounds.

Covalent Bonds

1

1.1 LEWIS STRUCTURES AND BONDING

In oversimplified terms, chemical bonds constitute the glue that holds atoms together to make molecules. For our purposes, we must remember that all elements tend to fill their outermost electronic shells so that they have the number of electrons present in the outer electronic shells of the noble gases helium, neon, argon, and so forth. With regard to the first shell, two electrons are required for the element to resemble helium. The Lewis structure for helium is :He. In Lewis structures, only the electrons in the outermost orbit are shown. Hence, the correct Lewis structures of :N̤̈e: and :Ä̤r: are identical as far as the dots are concerned, but we realize that, although not shown, neon has two electrons in its first shell and eight in its second and outermost shell, whereas argon has two electrons in its first shell, eight in its second shell, and eight in its third and outermost shell. Perhaps more simply, the dots for argon represent the electrons in its third shell, while the dots for neon represent the electrons in its second shell; in helium the two dots represent the electrons in its first shell. Except for helium, eight electrons are needed in the outermost shell. This is the basis of the *octet rule*: Elements tend to gain or lose electrons so that there are precisely eight electrons in their highest principal energy level. Of course for hydrogen the magic number of electrons is two.

In *ionic bonds* we visualize that electrons are actually given by one atom to another, and the bond or glue is simply the electrostatic attraction between oppositely charged atoms. In *covalent bonds* electrons are shared between atoms and a bond exists when two electrons (or a pair of electrons) are shared between atoms. The sharing allows each atom to have eight electrons in its outermost shell, counting the shared ones.

The Lewis structures for some nonmetal elements commonly found in organic compounds are

·H, ·C·, :N·, :P·, :O:, :S:, :F·, :Cl·, :Br·, :I·

To fill their outermost shells, H needs one additional electron, C needs four, N and P need three, O and S need two, while the halogens (F, Cl, Br, I) all need one extra electron. All of these elements can form ionic bonds as well as covalent bonds. Most bonds actually have a partial ionic and partial covalent character. Nevertheless, in organic chemistry most of the bonds are predominately covalent in character, and we will assume that a bond is the covalent type unless the ionic charges are shown.

1.1.1 Single Bonds

The octet rule can be useful in indicating possible bonds between atoms. Suppose you are given the information that a compound with an empirical formula of CH_4 exists. By applying the octet rule (plus the rule of two for hydrogen) the location of bonds between atoms can be deduced. A trial-and-error method should lead to the correct Lewis structure of methane:

```
   H
   ··
H : C : H
   ··
   H
```

Lewis structure of methane

The method may be more apparent if the elements are written singly and the electrons of the hydrogen atoms are represented by *x*'s instead of by dots (of course one electron is the same as any/other; so the practice of distinguishing between the electrons of different elements has *no significance in structures* of compounds, but it helps account for electrons:

ˣH + ˣH + ˣH + ˣH + ·C·

Here, if a hydrogen atom could share one of the carbon's electrons, that would satisfy *its* needs. Since there are four hydrogen atoms and the carbon atom has four electrons to share, everything balances:

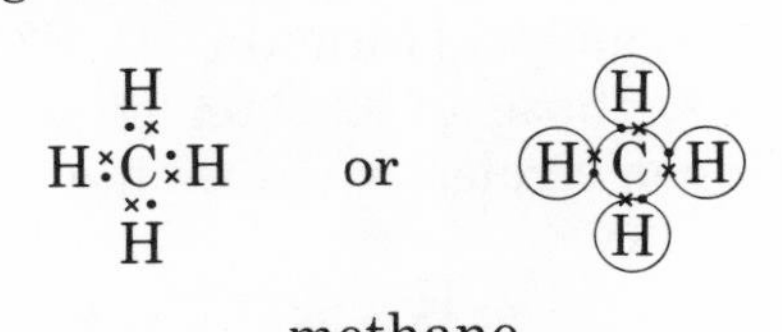

methane

Now the carbon atom has eight electrons "surrounding" it and each hydrogen atom shares two electrons. All atoms are satisfied. In the Lewis structure an electron pair is shown to one side of a hydrogen atom, but in molecules the electron pair circulates completely around the hydrogen atom's nucleus. Sometimes circles are added to Lewis structures to signify the orbital nature of the electrons involved in bonding. However, orbitals are not necessarily circular. If all atoms are not satisfied the structure is not correct because such a compound would not be stable. A methyl radical has an empirical formula of CH_3, but it is not stable, that is, it does not exist by itself. The hydrogen atoms are satisfied but the carbon atom is not.

```
  H
  ••
H:C:H
  •
```

As mentioned earlier, the sharing of a pair of electrons is the basis of a covalent bond. Rather than drawing the complete Lewis structure for a compound, covalent bonds are indicated by dashes. A dash represents the sharing of a pair of electrons between two atoms. Hence methane may be written as

```
    H
    |
H—C—H
    |
    H
```

methane

To be satisfied, a carbon atom must have four bonds since it will always need four additional electrons to fill its outermost shell. Yes, *carbon always has four bonds.* Knowing that carbon has four bonds, writing methane can be simplified further by writing the atoms associated with each carbon atom:

CH_4

methane

As shown here, four hydrogen atoms are associated with one carbon atom. Since carbon always has four bonds there is only one possible structure—the one shown previously.

Could the octet rule for all atoms be satisfied in a compound with the empirical formula CH_3F? Let the electrons of hydrogen be *x*'s, those of carbon be dots, and those of fluorine be small open circles. Then we have

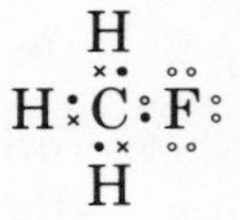

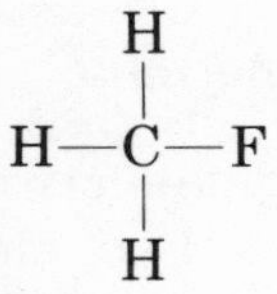

```
                       H
    H                  |
    ×• °°              |
  H×C•F°        H —— C —— F        or        CH3F
    •× °°              |
    H                  |
                       H
```

methyl fluoride

1.1.2 Double and Triple Bonds

Two atoms may share more than one pair of electrons. In fact, two pairs, or even three pairs, may be shared between the same two atoms. If only *one pair* is shared it is called a *single bond*. If *two pairs* are shared it is a *double bond*, and if *three pairs* are shared it is a *triple bond*. Ethylene, C_2H_4, is a compound that has a double bond between the two carbon atoms:

```
              H   H
  H H         |   |
  C::C   or   C = C
  H H         |   |
              H   H
```

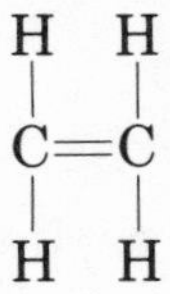

ethylene

Note that we draw a dash for each pair of electrons, and so two dashes represent a double bond. Acetylene, C_2H_2, has a triple bond between its two carbon atoms:

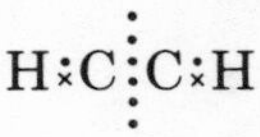

```
  H×C:::C×H     or     H—C≡C—H
```

acetylene

In all of the compounds the octet rule (the rule of two for hydrogen) is satisfied. Hydrogen can form only single bonds. What about oxygen? It needs two electrons to complete its outermost shell, and thus it can form double bonds. Consider formaldehyde, CH_2O, or HCHO. Since carbon needs four bonds and each hydrogen can provide only one, two bonds must be with oxygen:

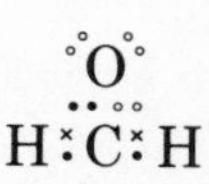

```
   °O°                  O
   ••°°                 ‖
  H×C×H      or     H — C — H
```

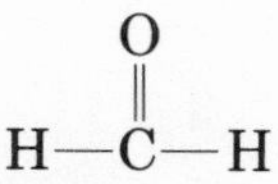

formaldehyde

Since an oxygen atom needs two electrons, oxygen should always form two bonds, but both bonds do not have to be with the same atom. In methanol, CH_4O, oxygen forms two single bonds, one with a hydrogen atom and one with a carbon atom:

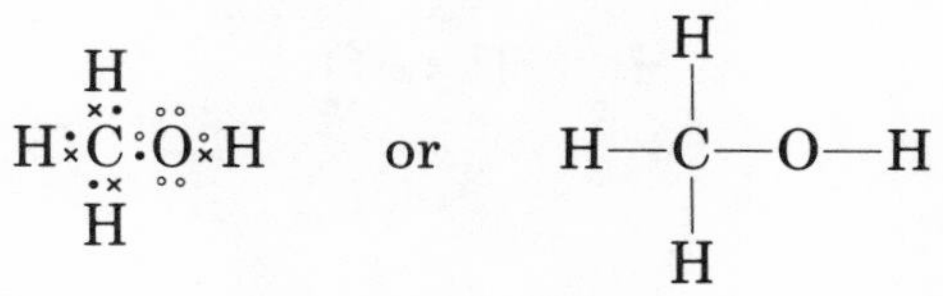

```
   H                    H
  ×· °°                 |
H×C:O×H      or     H—C—O—H
  ·× °°                 |
   H                    H
```

methanol

Nitrogen needs three electrons; so it must form three bonds. The Lewis structure of ammonia, NH_3, can be deduced easily:

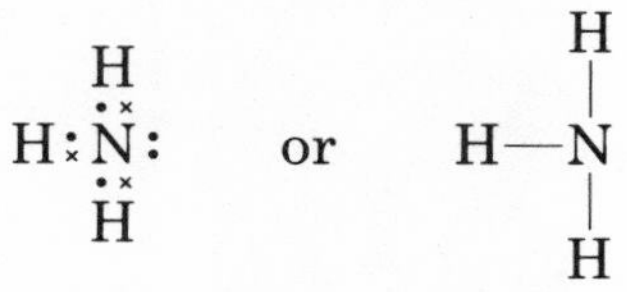

```
   H                H
   ·×               |
 H×N:       or   H—N
   ·×               |
   H                H
```

ammonia

Nitrogen needs three electrons to satisfy the octet rule, but the electrons may come from three different atoms, from two, or from only one. Therefore nitrogen can form single, double, and triple bonds. The structure of methylamine, CH_5N, shows single bonds only:

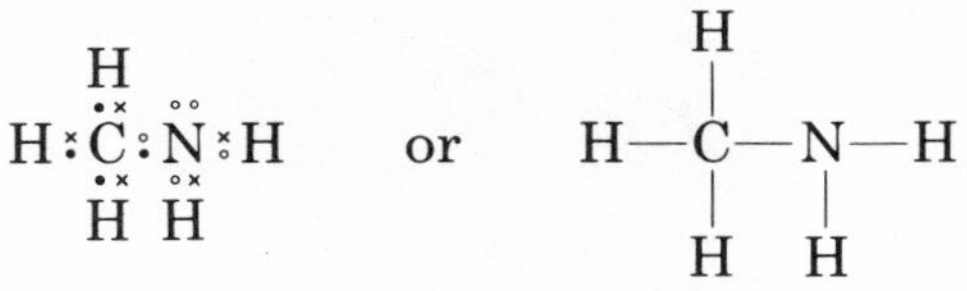

```
   H                    H
   ·× °°                |
 H:C:N×H      or    H—C—N—H
   ·× °×                |   |
   H  H                 H   H
```

methylamine

The structure of acetonitrile, C_2H_3N, shows a nitrogen atom with a triple bond:

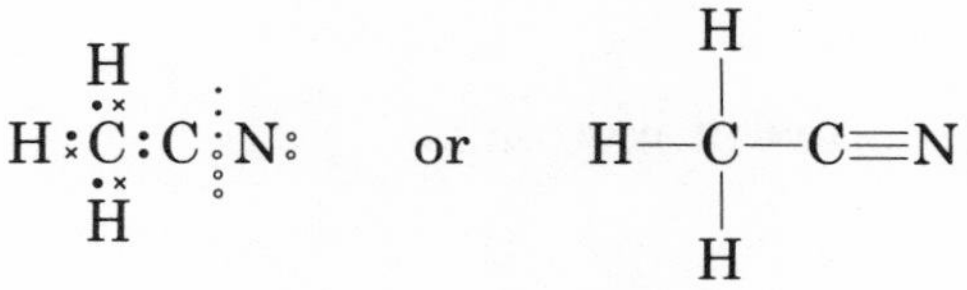

```
   H                    H
   ·×     ·             |
 H×C:C⋮N°      or   H—C—C≡N
   ·×     °             |
   H                    H
```

acetonitrile

1.1.3 Example

(a) By trial-and-error methods, determine a reasonable (octet rule satisfied) Lewis structure for the compounds with the following empirical formulas. (b) After obtaining the Lewis structures, use dashes to indicate bonds in the structures of the compounds.

Empirical formula: H_2O

$$\times H + \times H + \cdot\ddot{\underset{\cdot\cdot}{O}}\cdot$$

(a) H:Ö:H

(b) H—O—H

Empirical formula: $C_2H_4F_2$

$$2(\cdot\dot{\underset{\cdot}{C}}\cdot) + 4(H\times) + 2(:\ddot{\underset{\circ\circ}{F}}\cdot)$$

```
      H H
(a) :F:C:C:F:
      H H
```

```
        H   H
        |   |
(b) F — C — C — F
        |   |
        H   H
```

Actually, there is more than one correct answer in this last example.

```
      H H
(a) H:C:C:F:
      H:F:
```

```
        H   H
        |   |
(b) H — C — C — F
        |   |
        H   F
```

1.2 GEOMETRY OF THE BONDS FOR CARBON ATOMS

Review the structure of methane:

```
      H
      |
  H — C — H
      |
      H
```

methane

There are four hydrogen atoms bonded to a central carbon atom. *All four bonds are equivalent.* Therefore the angles of the bonds (lines between the atoms) should be equally spaced and be 109.5° apart. This yields a tetrahedron for methane with each hydrogen atom at a vertex and the carbon atom in the geometrical center, as shown in Fig. 1.1(a). A tetrahedron is a four-sided, triangular solid. Such a solid is generated if the lines between the hydrogen atoms represent edges of the triangular faces of the solid. A ball-and-stick model of

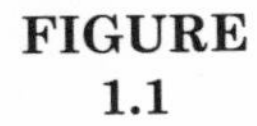

FIGURE 1.1

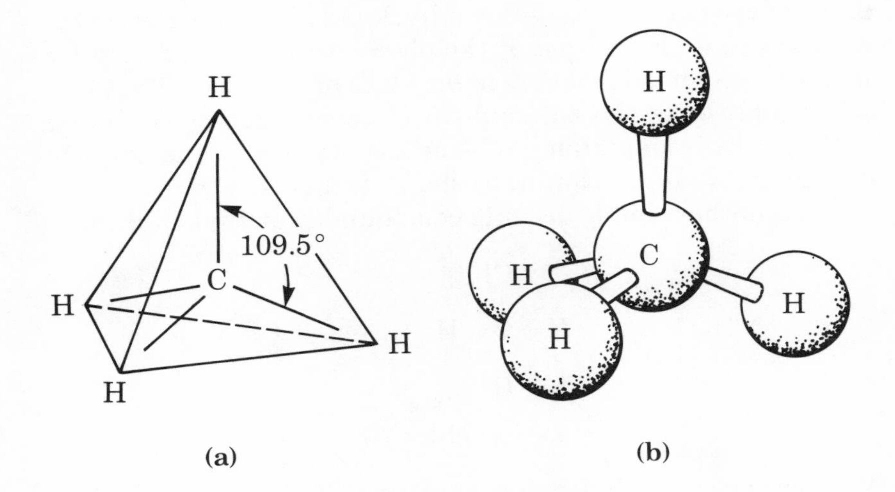

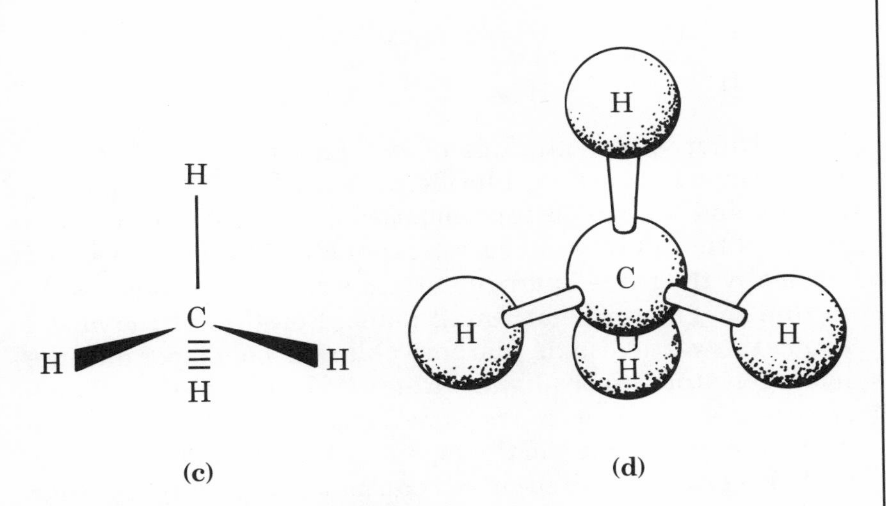

Representations of methane.

***(a)** A tetrahedron showing the carbon atom in the geometrical center and four bonds radiating to the hydrogen atoms at the vertexes of the tetrahedron.*

***(b)** A ball-and-stick model.*

***(c)** A projection structure, which is an approximation of the ball-and-stick model when oriented as shown in (d) or when viewed from one face of the tetrahedron shown in (a).*

***(d)** A ball-and-stick model showing the orientation of atoms in the projection structure of (c).*

methane, given in Fig. 1.1(b), may make this geometry clearer. In ball-and-stick models the balls represent atoms and the sticks represent bonds between atoms. Pay particular attention to Fig. 1.1(c). This is a projection structure used to illustrate the geometrical relationships shown in Fig. 1.1(d). Remember the following rules for projection structures. A *dashed* line stands for a bond that projects *back* away from the plane of the paper. A *heavy*, usually flared, line indicates bonds to atoms that are *in front* of the plane of the paper. A *regular* line indicates bonds that are *in the plane* of the paper. Thus Fig. 1.1(c) represents the ball-and-stick model when it is viewed with one hydrogen atom pointing directly away and slightly downward from the viewer, as shown in Fig. 1.1(d).

Another simple organic compound is methyl chloride:

```
  ..
 :Cl:
  ..                Cl
H:C:H               |
  ..             H—C—H        or      CH3Cl
  H                 |
                    H
```

CH_3Cl

methyl chloride

We may draw methyl chloride in any of the following four ways, but the actual structure of the molecule is the same:

```
   Cl          H           H            H
   |           |           |            |
H—C—H      H—C—Cl      H—C—H      Cl—C—H
   |           |           |            |
   H           H           Cl           H
```

Different orientations of ball-and-stick models of the same compound, methyl chloride, are shown in Fig. 1.2. Models (b), (c), and (d) can be superimposed on (a), as shown in the center of the figure. In all cases, if the chlorine atom is oriented vertically the three hydrogen atoms are turned concurrently so that they lie in a horizontal plane. Rotation of the model about the vertical axis (carbon–chlorine bond) permits the hydrogen atoms to be aligned. *Rule:* If by turning a ball-and-stick model, *one model can be superimposed on another*, then the models represent the same compound. You should experiment with this procedure to see if you can make two methyl chloride models that cannot be superimposed.

1.3 CARBON CHAINS

The number of compounds that could be formed from the common elements found in organic compounds is endless. Since one carbon atom can bond to another, and yet another,

FIGURE 1.2

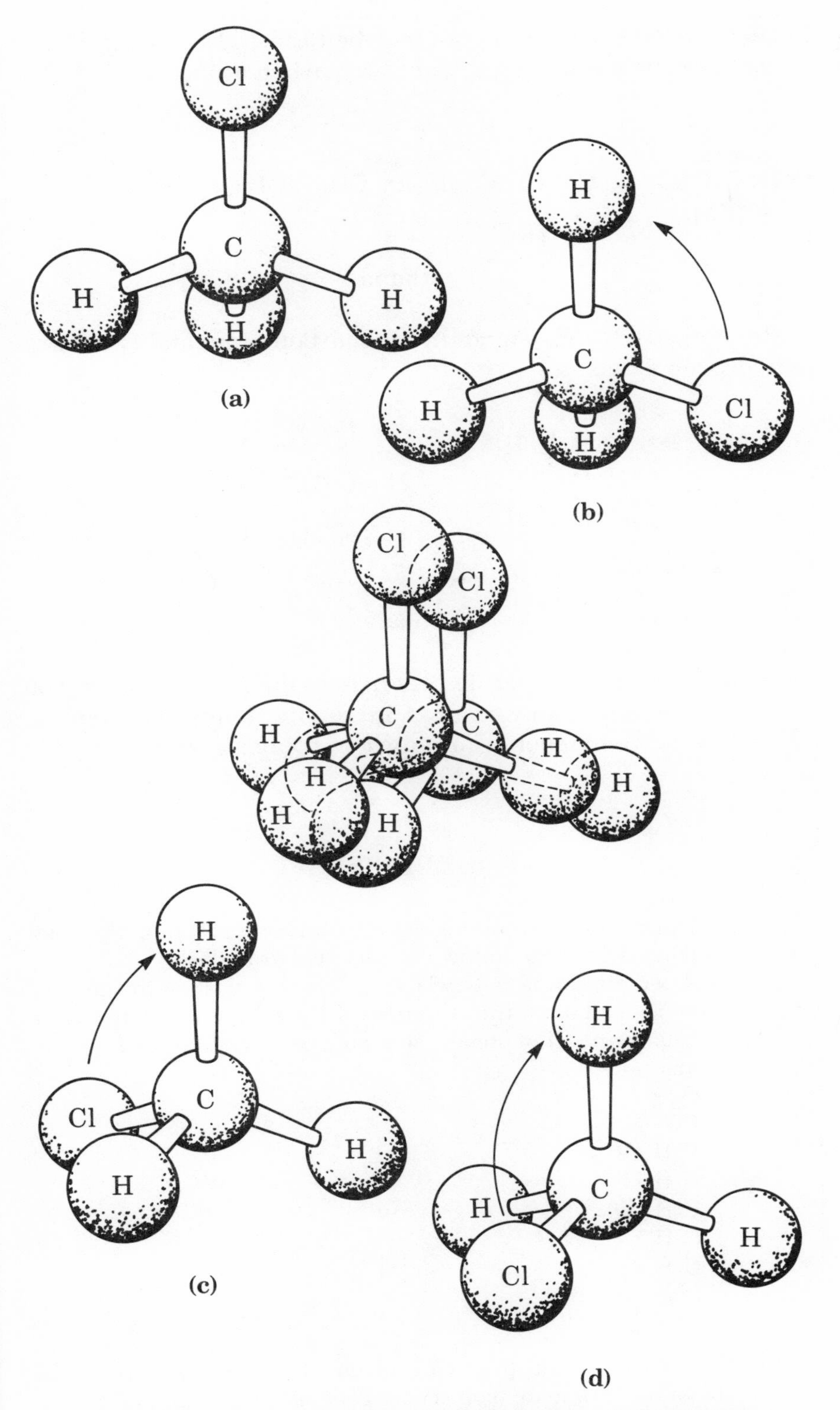

Orientations of models for methyl chloride. All models are identical and can be superimposed on one another. If the chlorine atom is oriented vertically, the hydrogen atoms are turned concurrently so that they lie in a horizontal plane. Then rotation of the model about the vertical axis (carbon–chlorine bond) permits the hydrogen atoms to be aligned.

carbon *chains* of any length may be formed. Ethane, C_2H_6, is the simplest chain having only two carbon atoms:

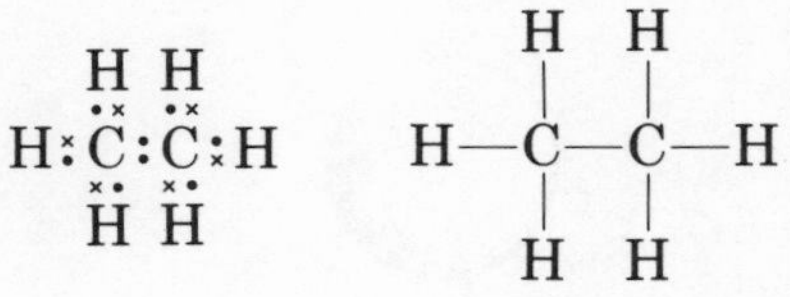

```
  H H            H   H
H:C:C:H       H—C—C—H       CH3—CH3     or     C2H6
  H H            H   H
```

ethane

Propane, C_3H_8, has an additional carbon atom and two more hydrogen atoms:

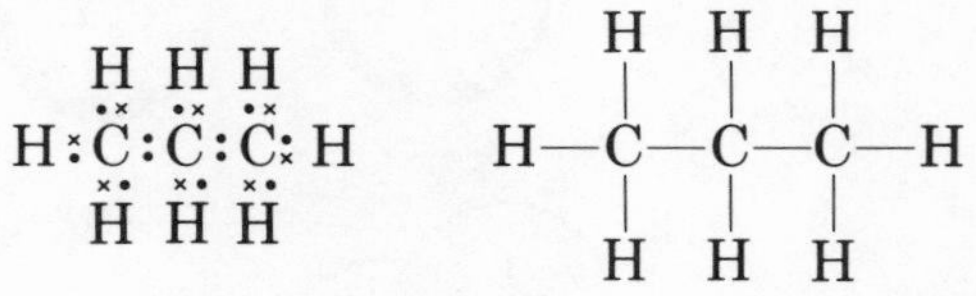

```
  H H H            H   H   H
H:C:C:C:H       H—C—C—C—H
  H H H            H   H   H

CH3—CH2—CH3      or      C3H8
```

propane

According to our basic rules, there is no limit on the number of carbon atoms that may occur in one chain. Hence the potential for an infinite number of organic compounds exists.

EXERCISES

1. **(a)** Draw one reasonable Lewis structure for each of the compounds with the following empirical formulas.
   (b) After obtaining a Lewis structure use dashes to indicate bonds and draw the structure of the compound in the more usual form. (For some, there may be more than one correct answer.)
   O_2
   CO_2
   C_2H_3Cl
   C_2H_7N
   C_2H_6
   C_3H_8
   $C_2H_3F_3$
   C_2H_4O
   $C_2H_4O_2$
2. When each of the following elements are components of a compound, how many bonds does each normally exhibit?
   (a) hydrogen
   (b) fluorine
   (c) oxygen
   (d) nitrogen
   (e) carbon

Alkanes, Alkenes, and Alkynes

2.1 ALKANES

Alkanes, *saturated hydrocarbons*, and *paraffins* are suitable names for this class of organic compounds containing only carbon and hydrogen. Alkanes have the general formula C_nH_{2n+2}, where n is any integer. Alkanes may be divided into (1) *normal alkanes*, whose carbon chains are *straight or unbranched*; and (2) *branched alkanes*. The normal alkanes form what is called a *homologous series*.

2.1.1 Homologous Series

The ability of carbon to form chains gives rise to a series of compounds differing by multiples of CH_2. The first 10 members of the homologous series of normal alkanes are given in Table 2.1. Each member of this potentially endless series is called a homolog of all other members. Homologs simply differ in the number of $\leftarrow CH_2 \rightarrow$-groups (called *methylene* groups) connected in an unbranched chain.

Compare the two condensed structural formulas given for butane in Table 2.1. In a structural formula, a set of parentheses with a subscript indicates that the group surrounded by parentheses is repeated the number of times shown by the subscript. Only the most condensed structural formulas for the homologs above pentane are given in Table 2.1, but their structures should be obvious at this point. The names in this series must be learned because the names of many organic compounds are derived from them.

2.1.2 Chain Isomers, or Skeletal Isomers

Because a carbon atom can form a bond with more than one other carbon atom, compounds may have branched chains. Branched alkanes may have the same empirical (or *most condensed*) formula as normal alkanes, but different struc-

TABLE 2.1
Names and Formulas for Some Alkanes

Alkane Name	Empirical Formula	Structural Formula	Condensed Structural Formula
Methane	CH_4	$\begin{matrix} & & H & & \\ & & \vert & & \\ H & - & C & - & H \\ & & \vert & & \\ & & H & & \end{matrix}$	CH_4
Ethane	C_2H_6	$\begin{matrix} & & H & & H & & \\ & & \vert & & \vert & & \\ H & - & C & - & C & - & H \\ & & \vert & & \vert & & \\ & & H & & H & & \end{matrix}$	CH_3-CH_3
Propane	C_3H_8	$\begin{matrix} & & H & & H & & H & & \\ & & \vert & & \vert & & \vert & & \\ H & - & C & - & C & - & C & - & H \\ & & \vert & & \vert & & \vert & & \\ & & H & & H & & H & & \end{matrix}$	$CH_3-CH_2-CH_3$
Butane	C_4H_{10}	$\begin{matrix} & & H & & H & & H & & H & & \\ & & \vert & & \vert & & \vert & & \vert & & \\ H & - & C & - & C & - & C & - & C & - & H \\ & & \vert & & \vert & & \vert & & \vert & & \\ & & H & & H & & H & & H & & \end{matrix}$	$CH_3-CH_2-CH_2-CH_3$ or $CH_3(CH_2)_2CH_3$
Pentane	C_5H_{12}	$\begin{matrix} & & H & & H & & H & & H & & H & & \\ & & \vert & & \vert & & \vert & & \vert & & \vert & & \\ H & - & C & - & C & - & C & - & C & - & C & - & H \\ & & \vert & & \vert & & \vert & & \vert & & \vert & & \\ & & H & & H & & H & & H & & H & & \end{matrix}$	$CH_3-CH_2-CH_2-CH_2-CH_3$ or $CH_3(CH_2)_3CH_3$
Hexane	C_6H_{14}		$CH_3(CH_2)_4CH_3$
Heptane	C_7H_{16}		$CH_3(CH_2)_5CH_3$
Octane	C_8H_{18}		$CH_3(CH_2)_6CH_3$
Nonane	C_9H_{20}		$CH_3(CH_2)_7CH_3$
Decane	$C_{10}H_{22}$		$CH_3(CH_2)_8CH_3$

tures. Consider butane, C_4H_{10}, for example:

$$\underset{\text{normal butane}}{CH_3{-}CH_2{-}CH_2{-}CH_3} \qquad \underset{\text{isobutane}}{CH_3{-}\overset{\overset{\displaystyle CH_3}{|}}{C}H{-}CH_3}$$

(Normal alkanes have been defined, but the word *normal* applies to other classes of organic compounds as well. The term *normal* always indicates a *straight* chain, and the name *normal butane* may be written as *n-butane.* It is important to remember that a small *n* stands for the word *normal.* As will be seen later, a capital *N* has a different meaning in naming compounds.) These two compounds are different and are isomers of each other. *Isomers are compounds that have the same empirical formula but different structures.* If *structures are truly different* the models of the compounds cannot be *superimposed* on one another. From Fig. 2.1 it is obvious that the models of normal butane and isobutane cannot be superimposed. Normal butane and isobutane are examples of *chain isomers*, or *skeletal isomers* (the carbon chain, or the carbon skeleton, of each compound differs). Except for the very simplest compounds, the empirical or most condensed formula

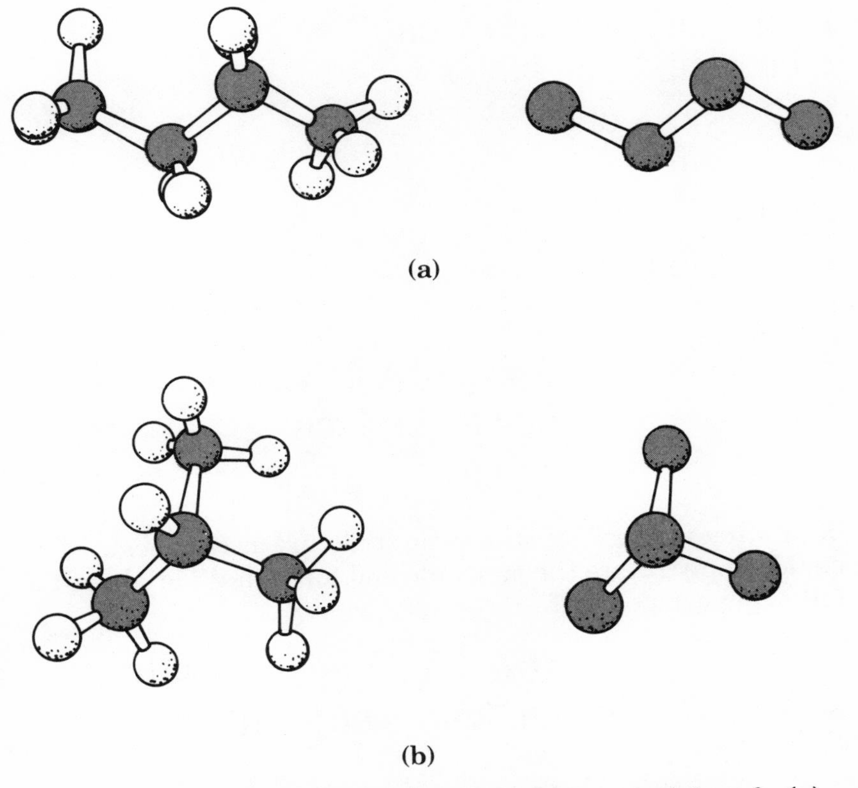

FIGURE 2.1

Complete models and carbon-skeleton models of (a) n-butane and (b) isobutane.

is ambiguous as far as the molecule's actual spatial configuration is concerned.

To determine if skeletal isomers are possible, look at the kinds of bonds that each carbon atom has. In *n*-butane two carbon atoms have two carbon neighbors, but each of the two end carbon atoms has only one carbon neighbor. In isobutane one carbon atom has three carbon neighbors while three carbon atoms are bonded to only one other carbon atom. Differences in structure may be seen more readily by drawing carbon skeletons in which all atoms except carbon are omitted intentionally. Of course, the rule of four bonds to a carbon atom appears to be violated in skeletons because only carbon–carbon (C–C) bonds are shown:

```
                      C
                      |
C—C—C—C             C—C—C
n-butane skeleton   isobutane skeleton
```

For butane there are two possible skeletal isomers, but the number of possible isomers increases rapidly as the number of carbon atoms in a molecule increases (see Table 2.2). The three isomers for the alkanes with an empirical formula of C_5H_{12} are:

```
1. CH3—CH2—CH2—CH2—CH3
2. CH3—CH2—CH—CH3
              |
              CH3

       CH3
       |
3. CH3—C—CH3
       |
       CH3
```

Would the compound

```
CH3—CH—CH2—CH3
    |
    CH3
```

be another isomer? It is the same structure as (2), because one could simply rotate the molecule end for end. What about the following compounds?

```
           CH3                      CH3
           |                        |
CH3—CH2—CH—CH3        CH3—CH2—CH
                                    |
                                    CH3
```

TABLE 2.2

Skeletal Isomers of Alkanes

Number of Carbon Atoms	Number of Possible Isomers
4	2
5	3
6	5
7	9
8	18
10	75
20	366,319
30	4,111,846,763

These are also the same as (2), because single C–C bonds are free to rotate, or we could rotate the molecules to look just like (2). To study bond rotation, ball-and-stick models are very useful, and you should therefore experiment with them. Remember the rule: If one model can be superimposed on another by rotating bonds or by turning the molecule, then the structures are identical and the compounds are not isomers of one another. The problem is that a tetrahedron cannot be drawn on paper easily, and so written structures do not reflect the three-dimensional picture. Isomer (2) could be drawn as

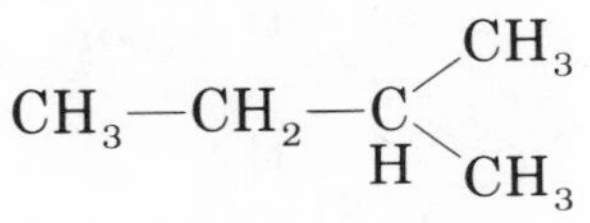

$$CH_3-CH_2-\overset{}{\underset{H}{C}}\genfrac{}{}{0pt}{}{\diagup CH_3}{\diagdown CH_3}$$

or even as

$$\begin{array}{ccc} & H & CH_3 \\ & | & \diagup \\ CH_2 & - C & \\ | & & \diagdown CH_3 \\ CH_3 & & \end{array}$$

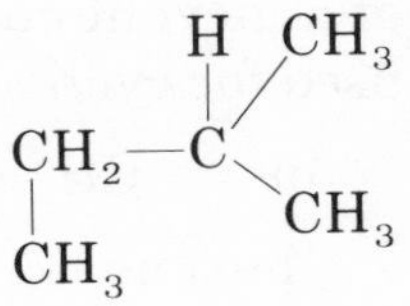

2.1.3 Cyclic Compounds

The ends of a carbon chain may be connected to form a ring. Because of the bond angles for carbon atoms, organic compounds frequently exist as five- or six-membered rings. Cyclohexane is an alkane that has six methylene groups ($-CH_2-$) in a ring. In the more complete structure, shown here on the left, all carbon and hydrogen atoms are given, but the bonds between the carbon atoms are emphasized:

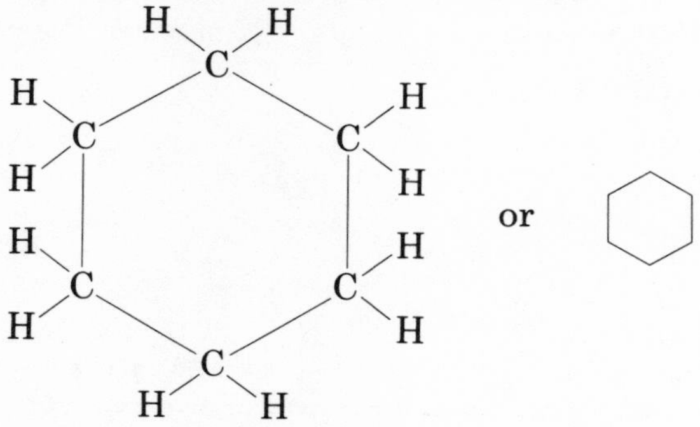

cyclohexane

In the simple structure on the right it is understood that a carbon atom is located at each vertex of the hexagon, and each C–C bond is drawn. Further, it is understood that unless double bonds are indicated or other groups are attached, *the remainder of the bonds from each carbon atom are with hydrogen atoms*. By using these rules, diagrams can be simplified greatly. However, the rings are not planar but rather are buckled as shown in Fig. 2.2.

Cyclopentane is an example of a five-membered ring:

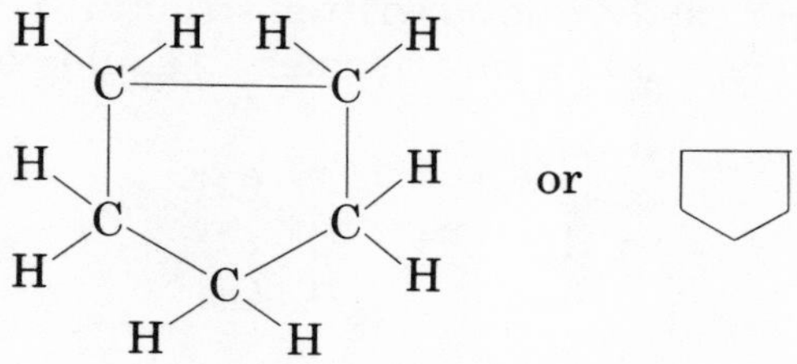

cyclopentane

Often structures of noncyclic compounds are shown in a manner similar to that used for cyclic compounds:

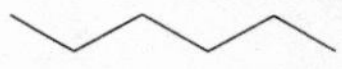

or $CH_3(CH_2)_4CH_3$

hexane

Because of the tetrahedral angles of carbon bonds, carbon skeletons are *not truly straight chains*. However, *n*-alkanes *can* have their C–C bonds rotated so that the *carbon skeleton* lies in a plane. This is best illustrated by making a long hydrocarbon with ball-and-stick models. When we extend the chain to its maximum length, the carbon atoms appear as a sawtooth curve (Fig. 2.3). The carbon skeleton is the basis for the highly simplified line structure just shown (on the left). Like the cyclic structures, each *vertex* (sawtooth) and each *end* represent a carbon atom *that has the remainder of its bonds saturated with hydrogen atoms*. Thus each end of the sawtooth chain stands for a CH_3. The *lines* simply *show the C–C bonds* in the molecule. No hydrogen atoms are shown, although it must be understood that they are there. If other groups are attached, they *are shown* attached to the vertex corresponding to the appropriate carbon atom:

$$
\text{(sawtooth structure with Cl on C2 and C3)} \quad \text{or} \quad \underset{}{CH_3}-CH_2-\underset{\displaystyle Cl}{\underset{|}{CH}}-\underset{\displaystyle Cl}{\underset{|}{CH}}-CH_3
$$

2,3-dichloropentane

In sawtooth structures groups radiate from the appropriate sawtooth points. That way, the structures are less crowded and show the true spacial orientation of the molecule more accurately.

FIGURE 2.2

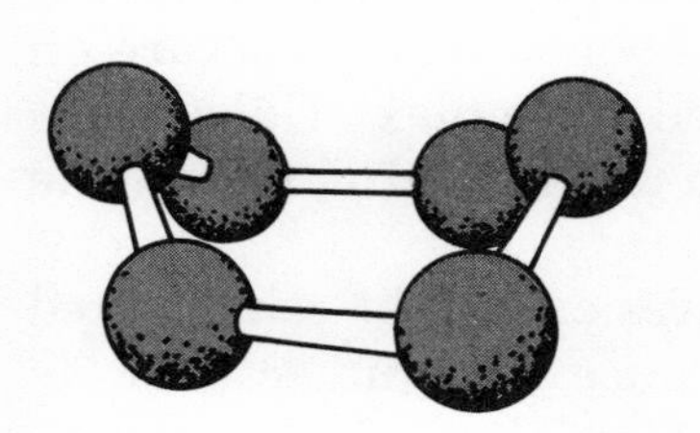

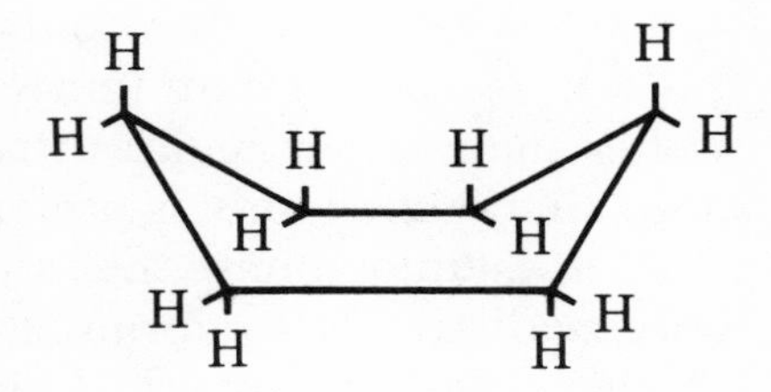

(a)

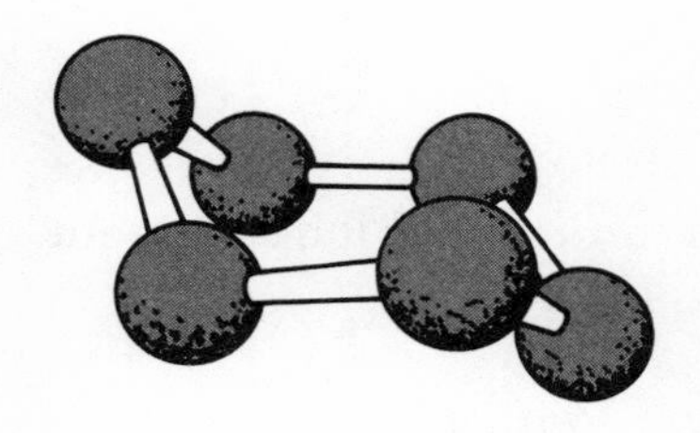

(b)

Models of the carbon skeletons and drawings of the (a) "boat" and (b) "chair" forms of cyclohexane, a six-membered cyclic alkane.

FIGURE 2.3

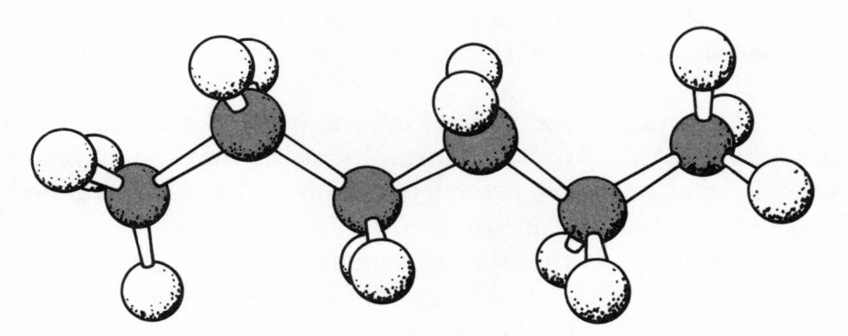

A ball-and-stick model of n-hexane, showing the sawtooth arrangement of the carbon skeleton.

Branched-chain alkanes, such as

$$CH_3—CH_2—\underset{\displaystyle CH_3}{\underset{|}{CH}}—CH_2—CH_2—CH_3$$

3-methylhexane

may be drawn as sawtooth structures in two ways:

(sawtooth structure with CH_3) or (sawtooth structure)

The structure on the right is the most abbreviated form and is a logical extension of the idea used to draw sawtooth structures—that is, show only C–C bonds. The structure on the left is correct but unnecessarily complex. Unless one wishes to draw particular attention to the methyl group, the more abbreviated form is preferred.

It is unfortunate that structures cannot be represented more realistically by formulas, because the connections between atoms determine the "topology" of the molecule, which is important in biological systems.

2.1.4 Alkyl Radicals

Reference is frequently made to alkyl *radicals*, which represent alkanes with a terminal hydrogen missing.

Alkane	*Alkyl Radical*
methane, CH_4	$CH_3—$, methyl
ethane, $CH_3—CH_3$	$CH_3—CH_2—$, ethyl
propane, $CH_3—CH_2—CH_3$	$CH_3—CH_2—CH_2—$, propyl
butane, $CH_3(CH_2)_2CH_3$	$CH_3(CH_2)_2CH_2—$, butyl
pentane, $CH_3(CH_2)_3CH_3$	$CH_3(CH_2)_3CH_2—$, pentyl
hexane, $CH_3(CH_2)_4CH_3$	$CH_3(CH_2)_4CH_2—$, hexyl
heptane, $CH_3(CH_2)_5CH_3$	$CH_3(CH_2)_5CH_2—$, heptyl
octane, $CH_3(CH_2)_6CH_3$	$CH_3(CH_2)_6CH_2—$, octyl

If R—H represents the normal alkane, then R represents the radical. An *R* in a structural formula means that some kind of radical—methyl, ethyl, propyl, butyl, or the like—is attached. In other words, it is an *unspecified alkyl radical.* Note that a radical's name is the alkane's name with a *-yl* substituted for the *-ane.* This also applies to the general name: Alk*ane* → alk*yl* radical or group.

Although it is not important for naming compounds, an unattached radical is called a *free radical.* As mentioned in

Section 1.1.1, radicals like the methyl radical ($CH_3\cdot$) are not complete molecules since they do not obey the rule of eight or two. However, they can be produced, but they are highly reactive and live only a short time. With its odd, unpaired electron, a free radical will combine with other atoms, such as a hydrogen atom ($H\cdot$), to form a stable chemical bond readily.

2.1.5 Functional Groups

An atom or groups of atoms bonded together within a molecule may be called a *functional group*, or more simply a *group*, if they are distinguishable from the remainder of the molecule. For example, two methyl groups (one at each end of the carbon chain) are present in *n*-butane. The word *functional* implies that the group may undergo chemical reactions readily when subjected to certain conditions.

2.1.6 Naming Alkanes

The names of all alkanes end with *-ane*. An example is *methane*. As indicated earlier, the names of the straight-chain alkanes are the basis for naming many other organic compounds. The names of the normal series, above butane, indicate the number of carbon atoms contained in each, and so learning them is fairly easy. How do we name members of the branched series such as

$$\begin{array}{l} CH_3—CH_2—CH_2—\underset{\displaystyle CH_3}{\underset{|}{CH}}—CH_3 \end{array}$$

We simply look for the longest straight chain and that becomes the "base" name. Here, there are five carbon atoms in a continuous chain; thus the base name is *pentane*. But there is a CH_3-group (called a methyl group, or a methyl radical) attached to one of the carbon atoms. The location of the methyl group is indicated by numbering the carbon atoms in the basic alkane, for example,

$$\overset{5}{CH_3}—\overset{4}{CH_2}—\overset{3}{CH_2}—\underset{\displaystyle CH_3}{\underset{|}{\overset{2}{CH}}}—\overset{1}{CH_3}$$

Then the appropriate number is placed before the group's name. Thus *2-methylpentane* indicates that a methyl group replaces a hydrogen atom on the #2 carbon atom of pentane.

The name of the following structure is also 2-methylpentane:

$$\begin{array}{l}CH_3-CH-CH_2-CH_2-CH_3\\ \qquad\;\;|\\ \qquad CH_3\end{array}$$

2-methylpentane
not
4-methylpentane

This structure could be turned end for end to make it look like the first structure. *Rule:* Numbers may proceed from either direction of the molecule, but one must strive to have the lowest numbers possible. Since both *2-methylpentane* and *4-methylpentane* describe the molecule, the first is correct because it uses a lower number.

Consider the compound

$$\begin{array}{l}CH_3-CH_2-CH-CH_3\\ \qquad\qquad\quad|\\ \qquad\qquad CH_2\\ \qquad\qquad\quad|\\ \qquad\qquad CH_3\end{array}$$

3-methylpentane

At first glance we might be inclined to name this compound *2-ethylbutane*, but that is not correct because the name should be based on the longest continuous chain in the molecule. Since single bonds are free to rotate, the structure could just as well have been written as

$$\begin{array}{l}CH_3-CH_2-CH-CH_2-CH_3\\ \qquad\qquad\quad|\\ \qquad\qquad CH_3\end{array}$$

Now it looks like *3-methylpentane*. An effort is made to draw structures in their most logical form, but this is not always possible.

$$\begin{array}{l}CH_3-CH-CH-CH_3\\ \qquad\;\;|\quad\;\;|\\ \qquad CH_3\; CH_3\end{array}$$

2,3-dimethylbutane

$$\begin{array}{l}\qquad\qquad\qquad CH_3\\ \qquad\qquad\qquad\;\;|\\ CH_3-CH_2-C-CH_3\\ \qquad\qquad\qquad\;\;|\\ \qquad\qquad\qquad CH_3\end{array}$$

2,2-dimethylbutane

These structures illustrate another feature of naming compounds. Two methyl groups are present in each compound. These are indicated by adding *di* before *-methyl-*. The prefixes used to indicate the various numbers of groups present in a molecule are given in Table 2.3. *One number must be used to* *indicate the position of each group.* That is, there must be *a number for each group even if the groups are attached to the same carbon atom.*

If you review the different structures used here as examples in naming alkanes you will see that four isomers of C_6H_{12} have been presented. The fifth possible isomer for that molecular formula is *n*-hexane.

The following structure illustrates how to name an alkane that has two or more different branches:

$$
\begin{array}{l}
CH_3-CH_2-CH_2-CH-CH-CH_2-CH_3 \\
\qquad\qquad\qquad\quad\;\, | \qquad\; | \\
\qquad\qquad\qquad\quad CH_2 \;\; CH_3 \\
\qquad\qquad\qquad\quad\;\, | \\
\qquad\qquad\qquad\quad CH_3
\end{array}
$$

4-ethyl-3-methylheptane

Numbers in a name are always separated from letters by a dash, and *the name of the last group or radical is joined to the base name as a continuous word.* The groups may be given in alphabetical order or in the order of decreasing complexity.

2.1.7 Uses of Alkanes

Petroleum is a mixture of alkanes, both normal and branched chains, with one to 40 carbon atoms. Petroleum may be fractionated (separated into fractions) by distillation. The most easily vaporized alkanes distill off at the lowest temperatures, and these are the short-chain compounds. Oil refineries perform the distillation process. Gasoline is a mixture of normal- and branched-chain alkanes that contain 4 to 12 carbon atoms. Diesel fuel is a fraction that distills at higher temperatures and consists of longer-chain hydrocarbons. Motor oils and lubricating greases are the very long chain

TABLE 2.3

Prefixes Used in Naming Organic Compounds

Prefix	Number of Groups
mono-	1
di-	2
tri-	3
tetra-	4
penta-	5
hexa-	6
hepta-	7
octa-	8
nona-	9
deca-	10

alkanes derived from petroleum. Paint thinner is another commonly used alkane mixture. Petroleum fractions are used to produce synthetic organic compounds. It may be surprising, but most of our pesticides, synthetic textile fibers, and plastics are made by series of organic reactions starting from petroleum fractions.

2.1.8 Properties of Alkanes

From your experience with the commonly used alkanes you know that, generally, they are not soluble in water. Because of this property, alkanes are said to be *hydrophobic* (*hydro* = "water," *phobic* = "being afraid"). *Hydrophilic* means "water loving," the opposite of hydrophobic. Alkanes will dissolve other hydrophobic alkanes or compounds; gasoline will dissolve grease.

As shown in Table 2.4 the boiling points of the different alkanes vary with the number of carbon atoms present. Methane (natural gas) is a gas at ordinary temperatures and pressure, as are ethane, propane, and butane. Pentane is a liquid, but very volatile. The boiling points of pentane, hexane, heptane, and so on, increase in order of the number of carbon atoms.

2.1.9 Reactions of Alkanes

Alkanes are not highly reactive compared to other organic compounds and most of the reactions used industrially are not comparable to reactions in living organisms. Therefore we will consider only the oxidation (burning) of alkanes. The balanced oxidation reactions for methane and ethane are given as

TABLE 2.4

Boiling Points of Some Alkanes

Alkane Name	Number of Carbon Atoms	Boiling Point (°C)
Methane	1	−164.0
Ethane	2	−88.6
Propane	3	−42.1
n-Butane	4	− 0.5
n-Pentane	5	36.1
n-Hexane	6	69.0
n-Heptane	7	98.4
n-Octane	8	125.7
n-Nonane	9	150.8
n-Decane	10	174.1

follows:

$$CH_4 + 2\,O_2 \longrightarrow CO_2 + 2\,H_2O$$

$$2\,(CH_3{-}CH_3) + 7\,O_2 \longrightarrow 4\,CO_2 + 6\,H_2O$$

Note that each carbon atom is combined with oxygen to form carbon dioxide, and all hydrogen atoms are combined with oxygen to form water. Considerable energy is released by these chemical reactions.

2.2 ALKENES

Hydrocarbons contain only carbon and hydrogen and may be further divided by distinguishing between those that have and those that do not have double or triple bonds. *Saturated* hydrocarbons (alkanes) *do not have any* double or triple bonds. *Unsaturated* hydrocarbons *do have* double or triple bonds.

Alkenes have at least one double bond in their molecules, as illustrated by the following:

$H_2C{=}CH_2$	$CH_3{-}CH_2{-}CH{=}CH_2$	$CH_3{-}CH{=}CH{-}CH_3$
ethylene	1-butene	2-butene

A general structure for alkenes is R—CH=CH—R. The R-groups may be the same or different alkyl radicals, or represent a hydrogen atom or even other unspecified groups.

2.2.1 Naming Alkenes

The alkenes 1-butene and 2-butene are isomers since they have the same molecular formula (C_4H_8) but *different structures*. The difference is the position of the double bond. To indicate the position of a double or triple bond, the bonds are numbered:

$$CH_3\overset{3}{-}CH_2\overset{2}{-}CH_2\overset{1}{-}CH_3$$

They can be numbered from either end of the chain, but again one should strive to have the lowest possible numbers in the name:

$$CH_2{=}CH{-}CH_2{-}CH_3$$

1-butene, *not* 3-butene

Although this structure looks different than that given previously for 1-butene, and might be called *3-butene*, remember that it could be turned end for end. Put another way, the numbering of bonds can proceed from either end of the carbon chain because the molecule could be rotated. The alkane corresponding to butene is *butane*; hence that is the base name. The suffix *-ane* indicates a saturated hydrocarbon whereas the suffix *-ene* indicates an alkene. Thus the suffix *-ene* is substituted for the suffix *-ane* in the base name, *butane*, to give *butene*. The number preceding the word indicates the position of the double bond.

There may be more than one double bond in a compound, for example,

$$CH_2{=}CH{-}CH{=}CH_2$$

1,3-butadiene

Since there are two double bonds the suffix is *-diene* and the numbers preceding the words indicate the position of the two double bonds. In this example the appropriate prefix, *-di-*, is attached to *-ene* to form the correct suffix. Trienes, tetraenes, and so on, also exist.

Cyclohexene is an example of a cyclic alkene:

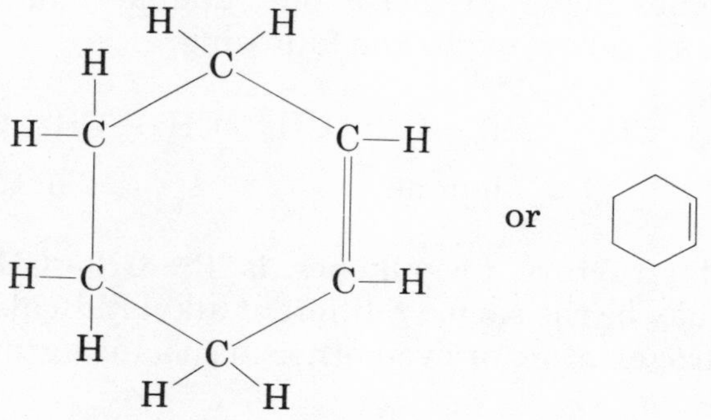

cyclohexene

As discussed previously, C–C bonds are drawn to obtain the abbreviated structure on the right. Since there are two bonds between two of the carbon atoms, both bonds are drawn. The same idea applies to sawtooth structures for open-chain compounds:

$$CH_3{-}CH_2{-}CH{=}CH{-}CH_2{-}CH_3$$ or

3-hexene

2.2.2 *Trivial and Systematic Names*

We may as well face another problem in naming organic compounds. There are two systems used: (1) the trivial or common names; and (2) the systematic, Geneva, or International Union of Pure and Applied Chemistry (IUPAC) names. In our discussion thus far, the second system has been used in describing how to name compounds. *Ethylene* is a common name; the corresponding systematic name is *ethene*. A number is not needed to indicate the position of the double bond because there is only one possibility. The trivial system lacks the logic of the systematic method, which was developed after many organic compounds had been discovered. In general, common and simple compounds are usually named by the trivial system whereas complex molecules are almost always named by the Geneva system, unless they are important molecules whose names are used frequently. *Vitamin* B_{12}, *ascorbic acid*, *penicillin K*, and *adenosine triphosphate* are some examples of commonly used trivial names of complex molecules. Either type of name may be considered to be correct.

2.2.3 Position Isomers

As mentioned before, 1-butene and 2-butene are isomers because they have the same molecular formula, C_4H_8, but they differ in the *position* of the double bond. Therefore we refer to them as *positional isomers*. A double bond may be called a functional group and the term *positional isomers* applies to other compounds that have different functional groups. As a general definition, *positional isomers are isomers among which the position of a particular functional group varies.*

2.2.4 Geometrical Isomers

Although rotation of single bonds is possible, *rotation is not possible for double or triple bonds*. This results in different structural features for compounds with double bonds. Consider the following structural formulas for 1,2-dichloroethene ($C_2H_2Cl_2$) and compare them to models shown in Fig. 2.4:

```
Cl      Cl          H       Cl
  \    /              \    /
   C=C                 C=C
  /    \              /    \
H       H          Cl       H

    or                  or

H       H          Cl       H
  \    /              \    /
   C=C                 C=C
  /    \              /    \
Cl      Cl          H       Cl
```

cis form of 1,2-dichloroethene — *trans* form of 1,2-dichloroethene

Since two points always lie in a plane, the two carbon atoms are in a single plane; thus by simply turning over the upper structure on the left, the lower-left structure would be obtained, and by turning over the upper structure on the right, the lower-right form would result. However, there is no way that a right form can be superimposed on a left form. Since the structures have the same molecular formulas, they fit the definition of isomers. If the two Cl-groups are on the *same side* of the molecule we say that they have the cis-*configuration*. If the two "like" groups are on *opposite sides* of the molecules we say that they have the trans-*configuration*. *Cis* and *trans* isomers are called geometrical isomers. "Like" groups might be hydrogen atoms, or any of the functional groups that will be discussed later.

Incidentally, yet another isomer of dichloroethene exists:

```
H         Cl
 \       /
  C==C
 /       \
H         Cl
```

1,1-dichloroethene

The empirical formulas of 1,1-dichloroethene, *cis*-1,2-dichloroethene, and *trans*-1,2-dichloroethene are the same, but all represent different structures. The 1,1-dichloroethene structure is a position isomer of the two geometrical isomers (*cis* and *trans* forms). This illustrates the idea that more than one type of isomer may be possible. There may be *many* points in a large molecule where isomerization is possible.

2.2.5 Properties of Alkenes

Alkenes are very similar to alkanes in their physical properties and are hydrophobic. In general, alkenes are soluble in alkanes and insoluble in hydrophilic solvents such as water. Compounds with double bonds have somewhat lower melting and boiling points compared to their analogous saturated compounds. For example, the boiling point of 1-hexene is 63.4°C compared to 69.0°C for n-hexane, and the melting points of these compounds are −139.5°C and −95.0°C respectively.

FIGURE 2.4

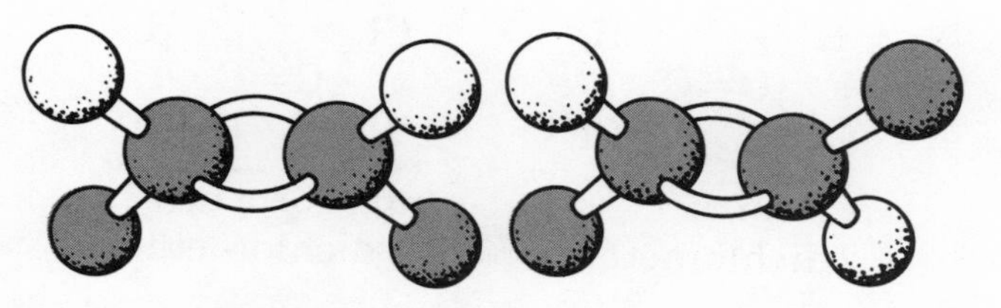

Models of 1,2-dichloroethene isomers. The cis isomer is on the left and the trans isomer is on the right.

2.2.6 Reactions of Alkenes

Alkenes are more reactive than alkanes. They may be hydrogenated (have H_2 added to them) to form alkanes:

$$H_2C{=}CH_2 + H_2 \longrightarrow H_3C{-}CH_3$$

ethylene or ethene → ethane

Although the details of reaction mechanisms will not be discussed here, it is convenient to think of reactions in a mechanistic fashion. For example, the preceding reaction can be described as "the addition of hydrogen across a double bond." Organic reactions are nothing more or less than a situation where the bonds between some atoms are broken and some new bonds are formed. Since a bond represents two shared electrons, we can visualize reactions as being the movement of pairs of electrons. A bond is *broken* when two atoms *stop sharing* a pair of electrons and a new bond is *formed* when two atoms *start sharing* a pair of electrons. An electron is an electron to an atom. It does not matter where the electron came from originally as long as it satisfies the atom's needs.

Consider the Lewis structure of H_2, hydrogen gas (H:H). Suppose that a molecule of H_2 comes close to a molecule of ethylene:

```
  H:H                 H—H
H:C⁞C:H      or    H—C=C—H
  Ḧ Ḧ                 |  |
                      H  H
```

Now suppose that the hydrogen atoms and carbon atoms "decide" to start sharing a pair of electrons:

```
  H:H              H H
H:C⁞C:H   ⟶     H:C:C:H
  Ḧ Ḧ              Ḧ Ḧ
```

The results are as follows. (1) The bond between the hydrogen atoms of H_2 is broken; hence those two atoms stop sharing electrons. (2) One of the double bonds between the carbon atoms is broken; and the carbon atoms stop sharing one pair, but continue to share the remaining pair, of electrons. (3) Single bonds are established between the carbon and hydrogen atoms described in (1) and (2); that is, the carbon atoms and "new" hydrogen atoms start sharing pairs of electrons. The product of the reaction, ethane, has the valences of all component atoms fully satisfied, and the same is true for the reactants, which were diatomic hydrogen and ethylene.

The addition of hydrogen is referred to as reduction. Thus the *hydrogenation* of ethylene means the same thing as the *reduction* of ethylene. In the preceding reaction ethylene was reduced to ethane, or we might say that the double bond has been reduced. *Reduction means the addition of electrons whereas oxidation means the removal of electrons.* These definitions apply in organic and biological chemistry as well as in inorganic chemistry; however, hydrogen atoms and oxygen atoms dominate the action in oxidation and reduction reactions. In the example given, we said that ethylene has been reduced, and it may be seen that two electrons (with their corresponding protons, the hydrogen nuclei) were added to ethylene. Consequently, saying that ethylene has been reduced is consistent with the usual definition of oxidations and reductions.

The reduction of ethylene illustrates the difference between saturated and unsaturated hydrocarbons. Ethane is saturated with hydrogen; no more can be added. Ethylene is an unsaturated hydrocarbon, and it can have hydrogen added to it by a reaction similar to the one given above.

In organic chemistry, *classes* of reactions exist. For example, consider the reduction (or hydrogenation) of an unspecified alkene. The reaction, as far as hydrogen and the double bond are concerned, is identical to that illustrated for ethylene. Hydrogen may be added across the double bond in both cases. Therefore a general reaction can be written:

$$\text{R—}\underset{\text{H}}{\underset{|}{\text{C}}}\text{=}\underset{\text{H}}{\underset{|}{\text{C}}}\text{—R} + \text{H}_2 \xrightarrow{\text{catalyst}} \text{R—}\underset{\text{H}}{\underset{|}{\overset{\text{H}}{\overset{|}{\text{C}}}}}\text{—}\underset{\text{H}}{\underset{|}{\overset{\text{H}}{\overset{|}{\text{C}}}}}\text{—R}$$

or

$$\text{R—CH=CH—R} + \text{H}_2 \xrightarrow{\text{catalyst}} \text{R—CH}_2\text{—CH}_2\text{—R}$$

From inorganic chemistry you will recall that a catalyst is a substance that increases the rate of a reaction but emerges from the reaction unchanged. Catalysts for hydrogenation reactions usually are finely divided metals or metal oxides such as Ni, Pt, or PtO_2.

In the preceding reaction R can be any radical, another type of group, or a hydrogen atom; and the R-groups may be the same or different. Remember, an R-group indicates that the group attached is unspecified. Hydrogenation of an alkene yields an alkane as illustrated in the general reaction. What if a molecule has more than one double bond; will the general reaction hold? Yes, every double bond should be expected to react, but of course one molecule of H_2 will be required for each:

$$CH_3{-}CH{=}CH{-}CH_2{-}CH{=}CH{-}CH_3 + 2H_2 \xrightarrow{\text{catalyst}} CH_3{-}CH_2{-}CH_2{-}CH_2{-}CH_2{-}CH_2{-}CH_3$$

Vegetable oils have long hydrocarbon chains with some unsaturated bonds (double bonds). Since alkanes have higher melting points than their corresponding alkenes, vegetable oils (liquid state) may be hydrogenated to yield fats (solid state). Crisco and Spry are commercial examples. You may have seen the phrase "partially hydrogenated corn oil" or "hydrogenated soybean oil" on food labels.

A diatomic halogen molecule (I_2, Cl_2, Br_2) can "add across a double bond." For example,

$$R{-}\underset{\substack{|\\H}}{C}{=}\underset{\substack{|\\H}}{C}{-}R + I_2 \longrightarrow R{-}\overset{\substack{I\\|}}{\underset{\substack{|\\H}}{C}}{-}\overset{\substack{I\\|}}{\underset{\substack{|\\H}}{C}}{-}R$$

This general type of reaction is used to measure the amount of unsaturated fatty acids in food oils. Since one molecule of I_2 is consumed for each double bond, the amount of I_2 taken up by a sample is directly proportional to the number of double bonds originally present in the sample. The addition of iodine to a double bond is called *iodination*, and the general term for the addition of any of the halogens is *halogenation*.

Learning general reactions like those illustrated here makes learning organic reactions easier. Instead of trying to remember all the reactions for all the possible alkenes, you should remember that hydrogen adds to double bonds and that all of the halogens add to double bonds. This little bit of knowledge will allow you to write reactions for any of the millions of alkenes that may exist.

For example, complete the following reaction:

$$CH_3{-}CH{=}CH{-}CH_2{-}CH_2{-}OH + I_2 \longrightarrow$$

You may wonder if the —OH-group on the end of the molecule will have any effect on the iodination of the double bond. Since the reactions of alcohols (OH-groups) have not been discussed, you cannot be expected to know that there will be no reaction involving the —OH-group. For the purposes of the present discussion assume that there is no reaction with other functional groups unless you have knowledge that indicates otherwise. Therefore the completed reaction should be

$$CH_3{-}CH{=}CH{-}CH_2{-}CH_2{-}OH + I_2 \longrightarrow CH_3{-}\underset{\displaystyle I}{\underset{|}{CH}}{-}\underset{\displaystyle I}{\underset{|}{CH}}{-}CH_2{-}CH_2{-}OH$$

Note that the reaction is balanced since it takes one molecule of diatomic iodine (I_2) for each double bond.

2.3 ALKYNES

Alkynes contain triple bonds and undergo reactions similar to those of the alkenes, but are uncommon in biological compounds. One example that may be of some interest is acetylene, the gas used with oxygen in torch welders. The burning (oxidation) of acetylene is the reaction

$$2\ (\underset{\text{acetylene}}{HC{\equiv}CH}) + 5\ O_2 \longrightarrow 4\ CO_2 + 2\ H_2O$$

Because of the heat evolved, it is called an exothermic reaction. Compare this reaction with those shown for the burning of methane and ethane. As with methane and ethane, all carbon atoms end up in CO_2 and all hydrogen atoms in H_2O.

The position of a triple bond is indicated by numbers, just as it is in alkenes, and *-yne* is substituted for *-ane* in the hydrocarbon name:

$$\underset{\text{1-butyne}}{CH_3{-}CH_2{-}C{\equiv}CH} \qquad \underset{\text{1,4-pentadiyne}}{CH{\equiv}C{-}CH_2{-}C{\equiv}CH}$$

EXERCISES

1. Draw the carbon skeleton for all of the possible skeletal isomers of pentane.
2. Name the following alkanes.

(a) 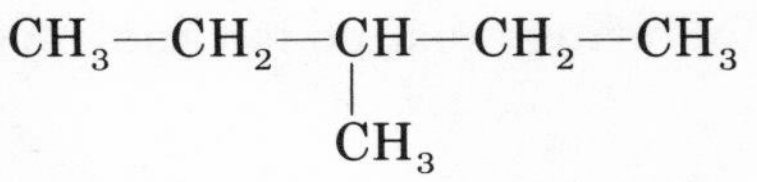 $$\begin{array}{l} CH_3-CH_2-\underset{\displaystyle CH_3}{\underset{|}{CH}}-CH_2-CH_3 \end{array}$$

(b) 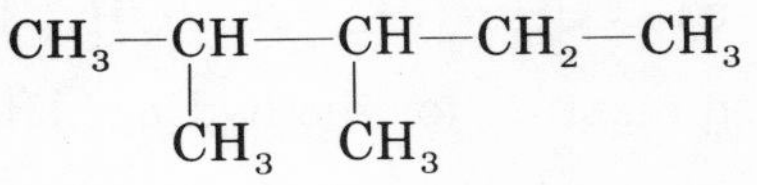 $$CH_3-\underset{\displaystyle CH_3}{\underset{|}{CH}}-\underset{\displaystyle CH_3}{\underset{|}{CH}}-CH_2-CH_3$$

(c) 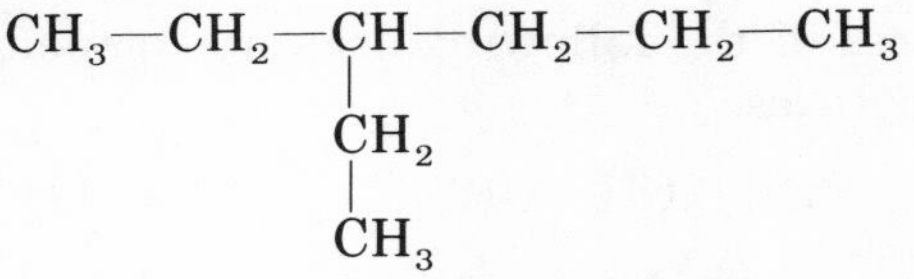 $$CH_3-CH_2-\underset{\displaystyle \underset{\displaystyle CH_3}{\underset{|}{CH_2}}}{\underset{|}{CH}}-CH_2-CH_2-CH_3$$

3. Indicate for each of the following structures whether the configuration of the double bond is *cis* or *trans*.

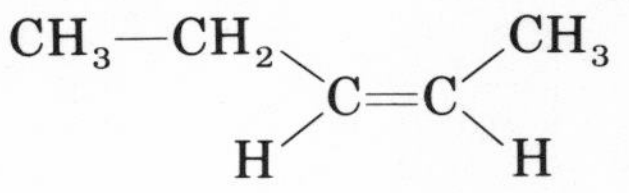

(a) 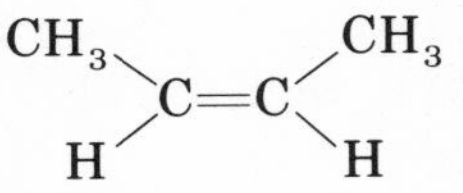 $$\begin{array}{ccc} CH_3 & & CH_3 \\ & C{=}C & \\ H & & H \end{array}$$

(b) $$\begin{array}{ccc} CH_3-CH_2 & & CH_3 \\ & C{=}C & \\ H & & H \end{array}$$

(c) $$\begin{array}{ccc} CH_3-CH_2 & & H \\ & C{=}C & \\ H & & CH_3 \end{array}$$

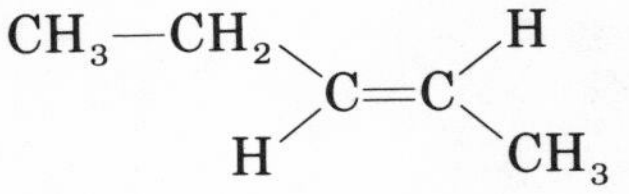

(d) 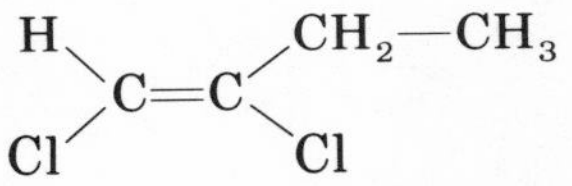 $$\begin{array}{ccc} H & & CH_2-CH_3 \\ & C{=}C & \\ Cl & & Cl \end{array}$$

(e) 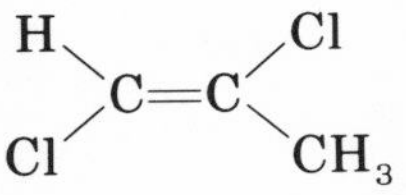 $$\begin{array}{ccc} H & & Cl \\ & C{=}C & \\ Cl & & CH_3 \end{array}$$

(f) 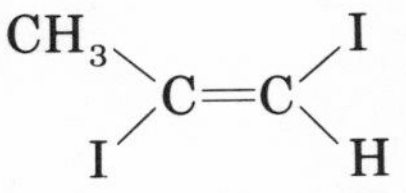 $$\begin{array}{ccc} CH_3 & & I \\ & C{=}C & \\ I & & H \end{array}$$

4. Complete and balance the following reaction for the oxidation of pentane.

$$CH_3{+}CH_2{+}_3CH_3 + O_2 \longrightarrow$$

5. Draw the structures for the following compounds.
(a) 2-methylheptane
(b) 3-ethyl-2,2-dimethylpentane
(c) 3-octene
(d) 1,4-decadiene

6. Name the following unsaturated hydrocarbons.
(a) $CH_3-CH_2-CH{=}CH-CH_3$
(b) $CH_3-CH{=}CH-CH{=}CH_2$
(c) $CH_3-CH_2-CH_2-CH_2-CH{=}CH-CH_3$
(d) $CH_3-C{\equiv}C-CH_2-CH_2-CH_3$

7. Complete and balance the following reaction for linoleic acid.

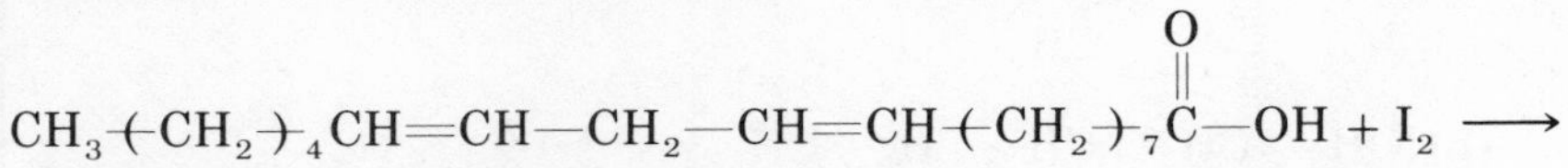

$$CH_3\!-\!(CH_2)_4CH{=}CH{-}CH_2{-}CH{=}CH\!-\!(CH_2)_7\overset{\overset{\displaystyle O}{\|}}{C}{-}OH + I_2 \longrightarrow$$

8. Show the complete balanced reaction for the hydrogenation of 2-pentene.

9. Convert the structures of the following organic compounds to line or sawtooth structures.

(a) 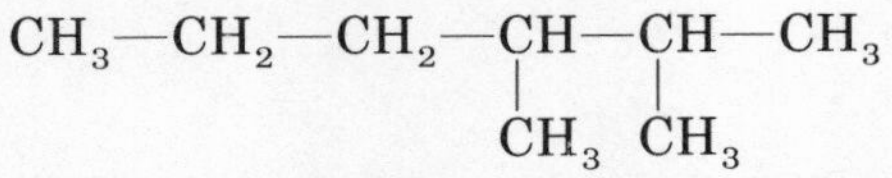 $CH_3{-}CH_2{-}CH_2{-}\underset{\displaystyle CH_3}{\underset{|}{CH}}{-}\underset{\displaystyle CH_3}{\underset{|}{CH}}{-}CH_3$

(b) 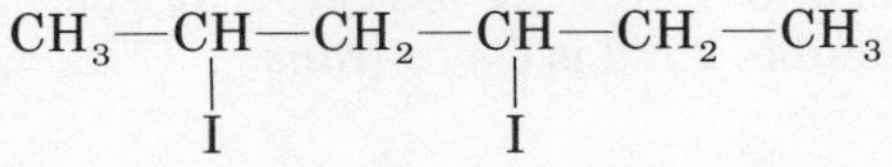 $CH_3{-}\underset{\displaystyle I}{\underset{|}{CH}}{-}CH_2{-}\underset{\displaystyle I}{\underset{|}{CH}}{-}CH_2{-}CH_3$

10. Supply the correct names for the following radicals.

(a) $CH_3{-}CH_2{-}$

(b) $CH_3\!-\!(CH_2)_4CH_2{-}$

(c) $CH_3\!-\!(CH_2)_5CH_2{-}$

(d) $CH_3\!-\!(CH_2)_8CH_2{-}$

Alcohols, Aldehydes, and Ketones

3.1 ALCOHOLS

Alcohols contain an —OH-group (hydroxyl group). Two very common examples are

CH_3—OH	CH_3—CH_2—OH
methyl alcohol or methanol	ethyl alcohol or ethanol

If the hydroxyl group is on the end of a carbon chain, the general formula for alcohols is R—OH. We can view an alcohol as an alkane with one of its hydrogen atoms replaced by a hydroxyl group or as water with one of its hydrogen atoms replaced by an alkyl radical, such as a methyl group:

```
    H             H
    |             |
H—C—H     H—C—O—H      H—O—H
    |             |
    H             H

 (R—H)       (R—O—H)       (H—O—H)
an alkane   an alcohol      water
```

3.1.1 Naming Isomers of Alcohols

Alcohols may be named by both the trivial and the Geneva systems (see Section 2.2.2). Trivial names are *methyl alcohol, ethyl alcohol, n-propyl alcohol, isopropyl alcohol*, and so on. In the Geneva system an *-ol* is substituted for the *-e* in the alkane's name. Hence we use *methanol, ethanol, propanol*, and so forth.

In propanol the —OH-group may be at the end or in the middle of the carbon chain,

$$CH_3—CH_2—CH_2—OH$$

propyl alcohol (trivial)
1-propanol (Geneva)

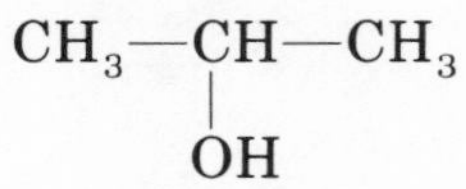

$$CH_3—\underset{\underset{OH}{|}}{CH}—CH_3$$

isopropyl alcohol (trivial)
2-propanol (Geneva)

so a method is needed to distinguish between these two positional isomers. In the trivial system the prefix *iso-* indicates that the —OH is not at the end of the chain; otherwise we would assume it was. In the Geneva system the *1-* or *2-* indicates the number of the carbon atom to which the —OH-group is attached. You might think that 3-propanol could exist, but if written, you will readily see that the structure could be rotated to give 1-propanol. Remember, when applying the Geneva system always use the lowest number possible.

Consider butanol. The Geneva-system name of each of the following isomers is given in parentheses under the trivial name:

$$CH_3—CH_2—CH_2—CH_2—OH$$ (*a primary alcohol*)

n-butyl alcohol
(1-butanol)

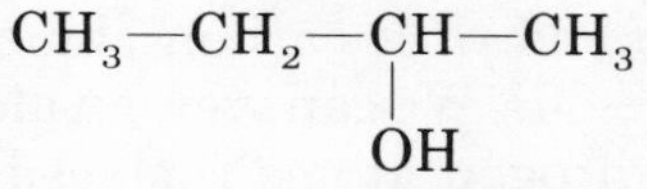

$$CH_3—CH_2—\underset{\underset{OH}{|}}{CH}—CH_3$$ (*a secondary alcohol*)

sec-butyl alcohol,
or
isobutyl alcohol
(2-butanol)

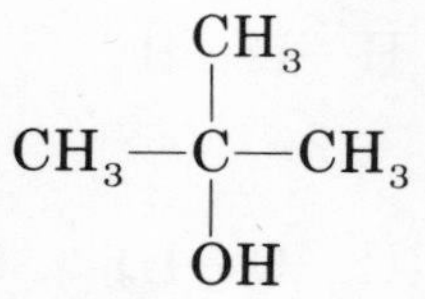

$$CH_3—\overset{\overset{CH_3}{|}}{\underset{\underset{OH}{|}}{C}}—CH_3$$ (*a tertiary alcohol*)

tert-butyl alcohol
(2-methyl-2-propanol)

Notice that in 1-butanol the hydroxyl group is attached to a terminal carbon atom, and it falls in the class called *primary (1°) alcohols*. In primary alcohols the hydroxyl is attached to a carbon atom, which in turn is attached to only *one* other carbon atom. All primary alcohols are 1-alkanols; that is, they have the hydroxyl group at one end of the carbon chain. R—OH is the general formula for primary alcohols. In 2-butanol the hydroxyl group is attached to a carbon, which in turn is attached to *two* other carbon atoms; hence it belongs to the class called *secondary (2°) alcohols*. The general formula for a secondary alcohol is

```
R—CH—R
   |
   OH
```

In 2-methyl-2-propanol the hydroxyl group is bonded to a carbon, which in turn is attached to *three* other carbon atoms, making it a *tertiary (3°) alcohol*. Tertiary alcohols have the general formula

```
     R
     |
  R—C—R
     |
     OH
```

Being able to distinguish between the classes of primary, secondary, and tertiary alcohols is important because the biochemical and organic reactions of these classes differ, as will be seen when aldehydes and ketones are discussed.

Note that in the Geneva system, *tert*-butyl alcohol does not have the word *butanol* in its name. Instead, 2-propanol is used as its base name, and the presence of the fourth carbon atom is indicated by adding *2-methyl-* as a prefix. Remember, the longest continuous chain of carbon atoms will form the alkane base name. To illustrate this point further, consider the following:

```
Five-carbon chain ←———————————————————————→
                                     OH
                                     |
                 CH3—CH2—CH2—C—CH3
                                     |
Six-carbon chain ←————————┐   CH2
                          |          |
                          ↓        CH3
```

3-methyl-3-hexanol,
not
2-ethyl-2-pentanol

Because single bonds can be rotated, this structure could have been written as

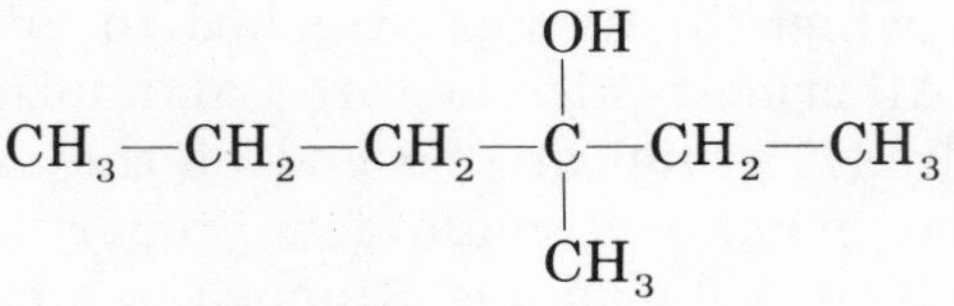

$$\begin{array}{c} \quad\quad\quad\quad\quad\quad\quad\quad\quad OH \\ \quad\quad\quad\quad\quad\quad\quad\quad\quad | \\ CH_3{-}CH_2{-}CH_2{-}C{-}CH_2{-}CH_3 \\ \quad\quad\quad\quad\quad\quad\quad\quad\quad | \\ \quad\quad\quad\quad\quad\quad\quad\quad\quad CH_3 \end{array}$$

3-methyl-3-hexanol

Now the correct name appears logical at first glance. For various reasons, the longest continuous chain of carbon atoms may not be written horizontally; therefore do not assume that it is.

3.1.2 Polyhydroxylic Alcohols

Alcohols may contain more than one hydroxyl group. Such alcohols are called polyhydroxylic alcohols, or polyols. Ethylene glycol, or antifreeze, is one example and it may be written in several ways:

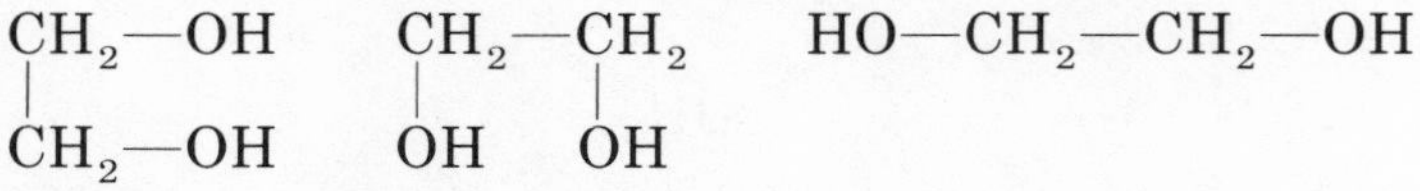

$$\begin{array}{l} CH_2{-}OH \\ | \\ CH_2{-}OH \end{array} \qquad \begin{array}{ll} CH_2{-} & CH_2 \\ | & | \\ OH & OH \end{array} \qquad HO{-}CH_2{-}CH_2{-}OH$$

ethylene glycol,
(1,2-ethandiol)

If there are two hydroxyl groups in a molecule the suffix is *-diol* instead of *-e* for the hydrocarbon name, and *-diol* instead of *-ol* for the same molecule with just one hydroxyl group. To indicate the position of the hydroxyl groups, the numbers of the carbon atoms to which the groups are attached are placed before the compound's name.

Glycerol (sometimes called glycerin) is a component of animal and plant fats:

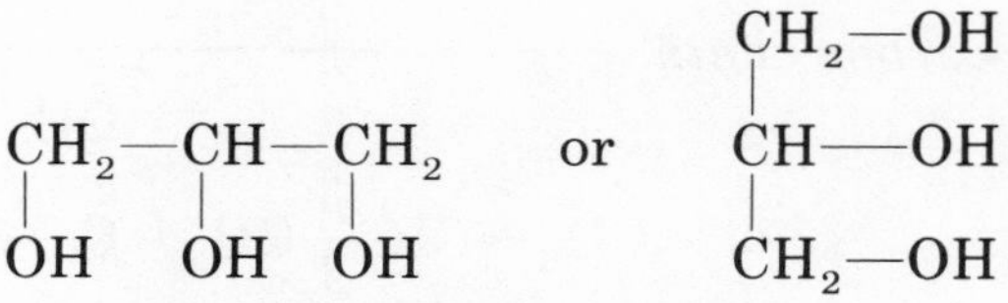

$$\begin{array}{lll} CH_2{-} & CH{-} & CH_2 \\ | & | & | \\ OH & OH & OH \end{array} \qquad \text{or} \qquad \begin{array}{l} CH_2{-}OH \\ | \\ CH{-}OH \\ | \\ CH_2{-}OH \end{array}$$

glycerol, or glycerin
(1,2,3-propantriol)

Glycerol is a triol. It has three hydroxyls attached to three different carbon atoms. The system can be extended to tetraols, pentaols, hexaols, and so on.

3.1.3 Naming an Alcohol with a Double Bond

How would you name an alkene that also has a hydroxyl group or an alcohol group? Instead of indicating the —OH-group by a suffix, we can indicate it by the prefix *hydroxy-*:

$$CH_3{-}CH_2{-}CH{=}CH{-}CH_2{-}OH$$

1-hydroxy-2-pentene

In naming organic compounds always look at the carbon skeleton first. In this case there are five carbon atoms in a straight line (no branches); therefore, if there were no functional groups, and if the compound were fully saturated, it would be *pentane*. This is the base name regardless of the functional groups attached. But there is a double bond at the #2 bond (only C–C bonds are counted, not the C–OH bond). So if there were no —OH, the compound would be *2-pentene*. A suffix has already been substituted for the *-ane* of the alkanes's name, and so another suffix cannot be added easily. Instead, *hydroxy-* is used as a prefix and a number is added to indicate the carbon atom to which the hydroxy group is attached. This yields the name *1-hydroxy-2-pentene*.

3.1.4 Properties of Alcohols

According to the Arrhenius theory, an *acid* is anything that will generate a higher concentration of hydrogen ions, $H^{\oplus}$ (more properly hydronium ions, $H_3^{\oplus}O$) than hydroxide ions, $OH^{\ominus}$, in water. For a *base* it is the reverse—more $OH^{\ominus}$ than $H^{\oplus}$. Water is neither an acid nor a base since its ionization produces equal concentrations of $H^{\oplus}$ and $OH^{\ominus}$:

$$H{-}O{-}H \rightleftharpoons H^{\oplus} + OH^{\ominus}$$

You might think that alcohols could behave as acids (donate $H^{\oplus}$) or bases (donate $OH^{\ominus}$) because of their similarity to water. But they do not:

$$R{-}O{-}H \not\rightleftharpoons RO^{\ominus} + H^{\oplus}$$

$$R{-}O{-}H \not\rightleftharpoons R^{\oplus} + OH^{\ominus}$$

They are surprisingly neutral compounds. Biochemically, alcohols are important and many of the biochemical reactions that they undergo are similar to organic reactions. Their reactions will be considered later.

We speak of alcohols or alcohol groups as being *polar*. To understand what is meant by such a statement, a review of the nature of water is necessary.

3.1.5 Water

Water is said to be a very polar molecule because the electrical charges are not evenly distributed among all atoms. Oxygen is highly *electronegative* (holds and attracts electrons very strongly compared to most other elements). Suppose that there are two positively charged bodies (nuclei) with two electrons between them. The two negative charges (electrons) will cancel the positive charges, making the combination neutral overall. However, *if the electrons are not shared equally between the positively charged atomic nuclei*, the area where the electrons are closer to one nucleus will have a slight negative charge whereas the area with the electrons somewhat removed from the nucleus will have a slight positive charge. Thus an unsymmetrical distribution of charges yields slightly more positive and slightly more negative areas within a body that is neutral overall. Less than one full electronic charge is indicated by the Greek letter δ.

In water, the electrons shared by the oxygen and hydrogen atoms are held more tightly by the oxygen. That is, compared to the size of the positive charge of its nucleus the outer (bonding) electrons are a little closer to the oxygen nucleus than they are to the hydrogen nuclei. Thus there is an unequal distribution of the charges.

One might expect water to be a linear molecule, but instead of the bonds being 180° apart they are 104.5° apart. Consequently the unequal distribution of charges within an irregularly shaped space yields a highly polar water molecule. Figure 3.1 shows an exaggerated Lewis structure for water and the general locations of the partial charges. The two slightly positive charges on the hydrogen atoms ($\delta^{\oplus}$) exactly cancel the single slightly negative charge ($\delta^{\ominus}$) on the oxygen atom.

The partial negative charge on the oxygen atom attracts the partial positive charge on one of the hydrogen atoms in a neighboring water molecule. Such attractions lead to hydrogen bonds. Consider the Lewis structures of two water mole-

FIGURE 3.1

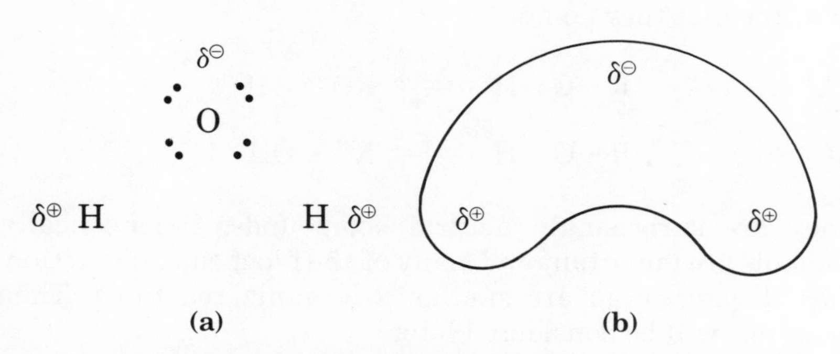

(a) The Lewis structure of water and the location of partial charges. (b) A general representation of the topology and charge distribution in a water molecule.

cules in Fig. 3.2(a). The partial charges have caused one hydrogen atom in one molecule to line up with the oxygen atom of the other molecule. Since that hydrogen atom is slightly deficient in electrons, it is attracted to the unshared electrons of the oxygen atom. From the Lewis structures one might be inclined to think that a bond has been established between these two atoms. This can occur, in which case the two electrons originally shared by the hydrogen atom must be left behind. The net result would be the formation of a hydronium ion, $H_3O^{\oplus}$, and a hydroxide ion, $OH^{\ominus}$. However, this situation represents one extreme; the other extreme comprises the original, separate, yet lined-up water molecules. Between these two extremes the hydrogen atom is being shared (unequally) between the oxygen atoms of the two water molecules. Of course, a hydrogen atom would be attracted more to the oxygen atom in its own molecule, but there would be some attraction to the oxygen atom in the neighboring molecule. This later, weaker attraction is called a *hydrogen bond.* A hydrogen bond is shown by a longer *dotted line* rather than by a solid dash, indicating that it is weak and easily broken compared to normal covalent bonds. The energy required to break the covalent bond between oxygen and hydrogen (O–H) is about 110 kcal/mol whereas the corresponding energy for a hydrogen bond (—O · · · H—) is about 4.5 kcal/mol. Nevertheless, hydrogen bonds have a profound effect on the structure of many biological compounds, because many such bonds may be involved in one molecule.

Previously, it was stated that a dashed line was used to indicate spatial arrangements in projection structures of molecules (see Fig. 1.1), and now a dotted line is being used to indicate a weak, hydrogen bond. Normally the similarity in

FIGURE 3.2

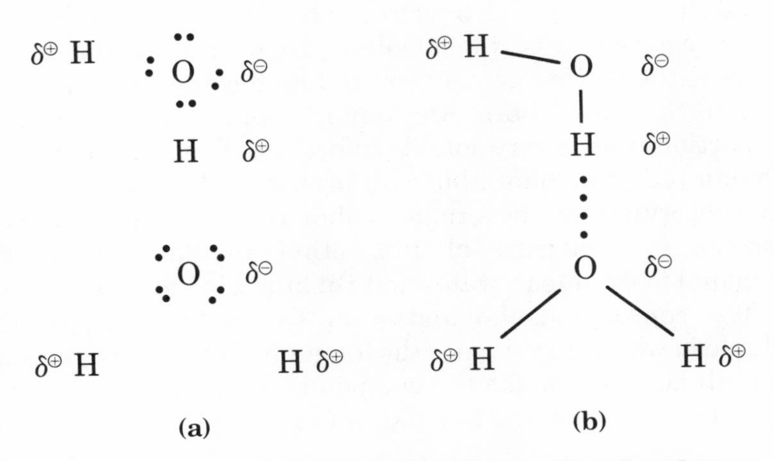

(a) The Lewis structure of two water molecules that are aligned to permit hydrogen bonding. (b) An illustration of hydrogen bonding (dotted line) between two water molecules.

drawings is not confusing because the rest of the structure indicates the type of bonds being illustrated. Generally, hydrogen bonds are longer than normal bonds and are illustrated that way as well.

Polarity of a compound (unequal charge distribution) is closely related to hydrogen bonding. Consider the alkanes. They are very nonpolar compounds: There is virtually no unequal charge distribution. The carbon atoms and hydrogen atoms, for example those in hexane, have electrons shared equally between the positive charges of the atoms' nuclei. Also there are no unshared electrons available in the outer orbitals. The conclusion here is that there is no hydrogen bonding in hexane and that the hydrogen atoms do not carry a significant partial positive charge ($\delta^{\oplus}$). Since the hydrogen and carbon atoms of hexane do not have partial charges, they would not be attracted by either the oxygen atoms or the hydrogen atoms of water molecules. This explains why hexane is not soluble in water.

There is a saying about solubilities: "*Like dissolves like.*" Pentane will dissolve in hexane in all proportions, and vice versa, and both are hydrophobic; pentane is like hexane. Water is not like hexane, and water does not dissolve (appreciably) in hexane. Consider methanol:

```
      H
      |
   H—C—O—H
      |
      H
```

methanol

Here an oxygen atom is present, and oxygen is "electron hungry." There is an unequal charge distribution between the O and the H of the hydroxyl group but it is not as great as in water. Nevertheless, the alcohol group is more like water whereas the methyl group is more like hexane. Which "like" wins in this case? Both. Methanol is soluble in water in all proportions and is very soluble in hexane. The same is true for ethanol (ethyl alcohol), but you probably already knew that from observing that beverage alcohol and water can be mixed. Gasohol is a mixture of 10% ethyl alcohol in gasoline. Propanol is soluble in water and rubbing alcohol is a mixture of 70% isopropyl alcohol and water. Butanol is only partially soluble in water. In this case the longer hydrophobic hydrocarbon tail begins to make the compound look and behave more like a hydrocarbon and less like water:

$$CH_3—CH_2—CH_2—CH_2—OH$$

hydrocarbon tail — polar head

The solubilities of higher homologs of alcohols in water decrease rapidly as the number of carbon atoms increases beyond four (butanol).

Is there hydrogen bonding between alcohol groups and water? Yes; it is similar to that in water but the hydrogen bonds are not as strong:

```
                     H
                    /
R—O—H····O
    ⋮              \
    H               H
     \
      O
     /
    H
```

hydrogen bonding between an alcohol and water molecules

Note that hydrogen bonding between alcohol molecules and water molecules can occur two ways: (1) A hydrogen atom of a water molecule may hydrogen bond with the oxygen atom of the alcohol and (2) the hydrogen atom of the alcohol's hydroxyl group may be hydrogen bonded to the oxygen atom of a water molecule. Hydrogen bonding between alcohol molecules also occurs, because the hydrogen atom of one molecule can hydrogen bond to the oxygen atom of a second molecule, or vice versa.

3.2 ALDEHYDES

Aldehydes have the general formula

```
     O
     ‖
R—C—H        or        RCHO
```

an aldehyde

In the formula to the right, the double bond from the carbon atom to the oxygen is not shown, but by convention it is understood that —CHO represents an aldehyde group. Therefore the functional group of an aldehyde may be represented as

```
   O
   ‖
—C—H        or        —CHO
```

an aldehyde group

Formaldehyde is the simplest of the aldehydes:

$$\mathrm{H{-}\overset{\overset{\displaystyle O}{\|}}{C}{-}H}$$

formaldehyde (trivial)
methanal (Geneva)

A 40% water solution of formaldehyde is called formalin and is used for preservation of biological materials and as an antiseptic. It has a vile odor and may be carcinogenic. Acetaldehyde is the next-to-the-simplest aldehyde:

$$\mathrm{CH_3{-}\overset{\overset{\displaystyle O}{\|}}{C}{-}H}$$

acetaldehyde (trivial)
ethanal (Geneva)

In the Geneva naming system the alkane with the same number of carbon atoms as the aldehyde to be named is the base name. Then the *-e* of the alkane's name is dropped and an *-al* is added. Hence:

$$\mathrm{CH_3{-}CH_2{-}\overset{\overset{\displaystyle O}{\|}}{C}{-}H}$$

propanal

Since an aldehyde group can exist only at the end of a chain, a number is not usually needed. *The carbon atom of the aldehyde group automatically becomes the #1 carbon atom of the compound.*

3.3 KETONES

The general formula for a ketone is

$$\mathrm{R{-}\overset{\overset{\displaystyle O}{\|}}{C}{-}R,\qquad R{-}CO{-}R\qquad or\qquad RCOR}$$

a ketone

Note that the *R*'s are used to denote alkyl radicals, which may be different or the same. For example, if the R-groups were methyl radicals the compound would be acetone:

$$CH_3-\overset{\overset{\displaystyle O}{\|}}{C}-CH_3$$

acetone (trivial)
propanone (Geneva)

The ketone group is represented as $-\overset{\overset{\displaystyle O}{\|}}{C}-$, sometimes referred to as a *keto* or a *carbonyl group*. In the Geneva system the position of the carbonyl group is indicated by the number of the carbon atom in the chain. In the example given here, there is only one position possible, and so a number is not needed. In the following examples numbers are needed. The correct suffix is *-one* (pronounced "own") and it replaces *-e* of the base alkane name:

$$CH_3-CH_2-CH_2-\overset{\overset{\displaystyle O}{\|}}{C}-CH_3 \qquad CH_3-CH_2-\overset{\overset{\displaystyle O}{\|}}{C}-CH_2-CH_3$$

2-pentanone   3-pentanone

Obviously, 2-pentanone and 3-pentanone are positional isomers. Again, the lowest numbers possible are chosen. Because of this rule, 4-pentanone would not be an acceptable name for 2-pentanone.

Diones and triones also exist:

$$CH_3-\overset{\overset{\displaystyle O}{\|}}{C}-CH_2-\overset{\overset{\displaystyle O}{\|}}{C}-CH_2-CH_3$$

2,4-hexandione

Ketones are quite similar to aldehydes with respect to the oxidation state of carbon. For our purposes, *oxidation* means the removal of hydrogen atoms, the addition of oxygen, or a combination of both. *Reduction* means the opposite—the addition of hydrogen and/or the removal of oxygen. As mentioned in Section 2.2.6, this is consistent with the definition of oxidation as the removal of electrons and reduction as the addition of electrons. A hydrogen atom is an electron donor, a reductant, or a reducing agent. An oxygen atom is an electron acceptor, an oxidant, or an oxidizing agent. Because of similar oxidation states for the carbon atoms, many reactions for aldehyde and ketone groups are also similar.

3.4 PREPARATION OF ALDEHYDES AND KETONES

The relationship between aldehydes and ketones becomes more apparent when the oxidation of alcohols is considered:

$$\mathrm{R{-}CH_2{-}OH} \xrightarrow[\text{conditions}]{\text{mild oxidizing}} \mathrm{R{-}\overset{\overset{\displaystyle O}{\|}}{C}{-}H}$$

a primary alcohol → *an aldehyde*

$$\mathrm{R{-}\overset{\overset{\displaystyle OH}{|}}{C}H{-}R} \xrightarrow[\text{conditions}]{\text{oxidizing}} \mathrm{R{-}\overset{\overset{\displaystyle O}{\|}}{C}{-}R}$$

a secondary alcohol → *a ketone*

Aldehydes are produced by oxidizing primary alcohols, and ketones are produced by oxidizing secondary alcohols. Hence the oxidation state of the carbon atom is the same for aldehydes and ketones. Aldehydes and ketones represent a higher oxidation state (for the carbon atom) than alcohols, and alcohols represent a higher oxidation state (of carbon) than alkanes. In each of the two reactions shown here, two hydrogen atoms have been removed from each alcohol to give the "oxidized" product (aldehyde or ketone). Note that the two reactions are not balanced with respect to hydrogen. The oxidizing agent used would accept the hydrogen atoms to become a reduced product or products, but without identifying the oxidizing agent the reduced product cannot be determined. Why would the oxidizing agent not be identified? First, several different reagents may perform the oxidation of the alcohols quite properly, and so their omission extends the idea of general reactions. Second, the emphasis is on the interconversion of organic compounds, not reactions of inorganic oxidizing agents such as $H_2Cr_2O_7$.

Sometimes [H] or [O] may be shown for oxidation-reduction reactions to indicate that an unspecified oxidizing or reducing agent participates in the reaction being illustrated. For example, the first reaction could have been written as

$$\mathrm{R{-}CH_2{-}OH + [O] \longrightarrow R{-}\overset{\overset{\displaystyle O}{\|}}{C}{-}H + HOH}$$

Again the oxidizing agent is not specified; instead it is represented as an active oxygen atom, which does not exist by itself. Although the latter reaction does not provide additional information, it is balanced.

3.5 REACTIONS OF ALDEHYDES AND KETONES

Aldehydes and ketones may be reduced to alcohols, which is the exact reverse of the oxidations just shown. In the reduction we can visualize an H_2 molecule being added across the double bond between the carbon and oxygen atoms of the aldehyde or keto group. This reaction is very similar to the one described for the reduction or hydrogenation of alkenes. Although the reductions of aldehydes and ketones are general reactions, they are illustrated here with specific compounds:

$$\underset{\text{acetaldehyde}}{CH_3—\overset{\overset{\displaystyle O}{\|}}{C}—H} + H_2 \xrightarrow{\text{catalyst}} \underset{\text{ethyl alcohol}}{CH_3—\overset{\overset{\displaystyle OH}{|}}{\underset{\underset{\displaystyle H}{|}}{C}}—H \text{ or } CH_3—CH_2—OH}$$

$$\underset{\text{acetone}}{CH_3—\overset{\overset{\displaystyle O}{\|}}{C}—CH_3} + H_2 \xrightarrow{\text{catalyst}} \underset{\text{isopropyl alcohol}}{CH_3—\overset{\overset{\displaystyle OH}{|}}{\underset{\underset{\displaystyle H}{|}}{C}}—CH_3}$$

Most aldehydes react with alcohols to form *hemiacetals*. Acid ($H^{\oplus}$) catalyzes the reaction:

$$\underset{\textit{an aldehyde}}{R—\overset{\overset{\displaystyle O}{\|}}{C}—H} + \underset{\textit{an alcohol}}{H—O—R'} \overset{H^{\oplus}}{\rightleftharpoons} \underset{\textit{a hemiacetal}}{R—\overset{\overset{\displaystyle OH}{|}}{\underset{\underset{\underset{\underset{\displaystyle R'}{|}}{\displaystyle O}}{|}}{C}}—H}$$

Ketones may react similarly:

$$\underset{\textit{a ketone}}{R—\overset{\overset{\displaystyle O}{\|}}{C}—R} + \underset{\textit{an alcohol}}{H—O—R'} \overset{H^{\oplus}}{\rightleftharpoons} \underset{\textit{a hemiketal}}{R—\overset{\overset{\displaystyle OH}{|}}{\underset{\underset{\underset{\underset{\displaystyle R'}{|}}{\displaystyle O}}{|}}{C}}—R}$$

The *hemi-* indicates that something is only half of what it could be.

A hemiacetal molecule may react with another molecule of an alcohol to form an *acetal*; a similar situation exists for hemiketals:

$$
\begin{array}{c}
\text{OH} \\
| \\
\text{R—C—H} \\
| \\
\text{O} \\
| \\
\text{R}'
\end{array}
+ \text{H—O—R}'' \underset{}{\overset{H^{\oplus}}{\rightleftharpoons}}
\begin{array}{c}
\text{R}'' \\
| \\
\text{O} \\
| \\
\text{R—C—H} \\
| \\
\text{O} \\
| \\
\text{R}'
\end{array}
+ H_2O
$$

a hemiacetal *an alcohol* *an acetal*

$$
\begin{array}{c}
\text{OH} \\
| \\
\text{R—C—R} \\
| \\
\text{O} \\
| \\
\text{R}'
\end{array}
+ \text{H—O—R}'' \underset{}{\overset{H^{\oplus}}{\rightleftharpoons}}
\begin{array}{c}
\text{R}'' \\
| \\
\text{O} \\
| \\
\text{R—C—R} \\
| \\
\text{O} \\
| \\
\text{R}'
\end{array}
+ H_2O
$$

a hemiketal *an alcohol* *a ketal*

Note that in these reactions the alkyl radicals are indicated as R′ or R″. This distinguishes them from the alkyl radicals of the aldehyde or ketone. So, R, R′, R″, R^1, R^2, R^3, and so on, in an organic structure simply distinguish between unspecified radicals in different positions.

Also shown in the preceding reactions are double arrows pointing in opposite directions. These indicate a "reversible" reaction. All reactions are reversible, but a more practical approach to the concept of reversibility is frequently taken. Consider the following reaction:

$$A \overset{\text{(theoretically reversible)}}{\rightleftharpoons} B$$

Assume that equilibrium can be achieved in a reasonable length of time (less than a week). If the equilibrium constant (K_{eq}) were large, say more than 1000, the reaction could be used to produce B from A, but not A from B, since the ratio $[B]/[A] = 1000/1$ at equilibrium. Hence we would call this an "irreversible" reaction and indicate that designation by showing a single arrow: $A \longrightarrow B$.

On the other hand, if the equilibrium constant were within a couple of orders of magnitude of 1.0 and equilibrium could be achieved readily, it might be practical to make *B* from *A* or vice versa. Many times utilization of the laws of mass action (adding or removing a reaction component) will permit us to obtain the desired product. Generally, we would call such a reaction a "reversible" reaction and indicate this by double arrows: $A \rightleftharpoons B$.

Double arrows have been used to indicate that the formation of hemiacetals, acetals, hemiketals, and ketals are reversible reactions, generally. The direction of reactions depends on the nature of the aldehydes or ketones and the alcohols. Hence some aldehydes form acetals readily whereas others do not. Sugars, such as glucose, form hemiacetals readily.

Glucose is a primary source of metabolic energy in almost all living organisms and it is a polyhydroxylic alcohol as well as an aldehyde:

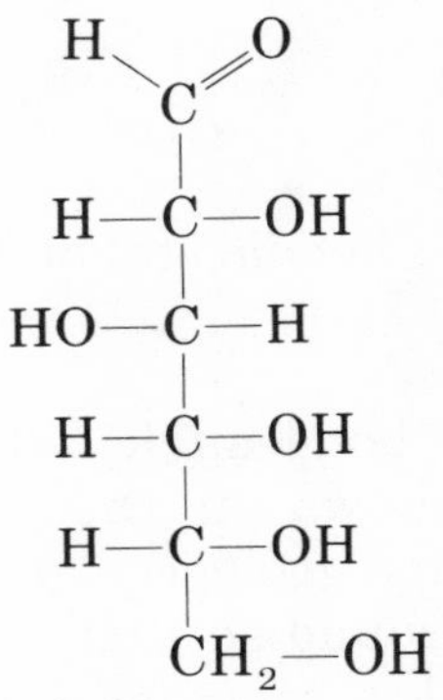

```
H     O
 \  //
   C
   |
H—C—OH
   |
HO—C—H
   |
H—C—OH
   |
H—C—OH
   |
  CH2—OH
```

glucose (trivial)
2,3,4,5,6-pentahydroxyhexanal (Geneva)

Since a hemiacetal is formed when aldehyde and alcohol groups react and both groups are present in the glucose molecule, it is reasonable to assume that such a reaction could occur between two glucose molecules or between *the two groups in the same molecule*. As mentioned earlier, carbon-containing compounds have a tendency to form five- or six-membered rings because of the geometry of carbon bonds. This helps to account for the fact that hemiacetal formation occurs *within the same molecule* for glucose. The aldehyde group (automatically the #1 carbon atom) reacts with the hydroxyl group on carbon #5 to form a six-membered ring that contains one oxygen atom. Perhaps the reaction would be visualized more easily if it were illustrated with 5-hydroxyhexanal, which would be like glucose except that all groups that do not participate in the reaction have been replaced by hydrogen atoms:

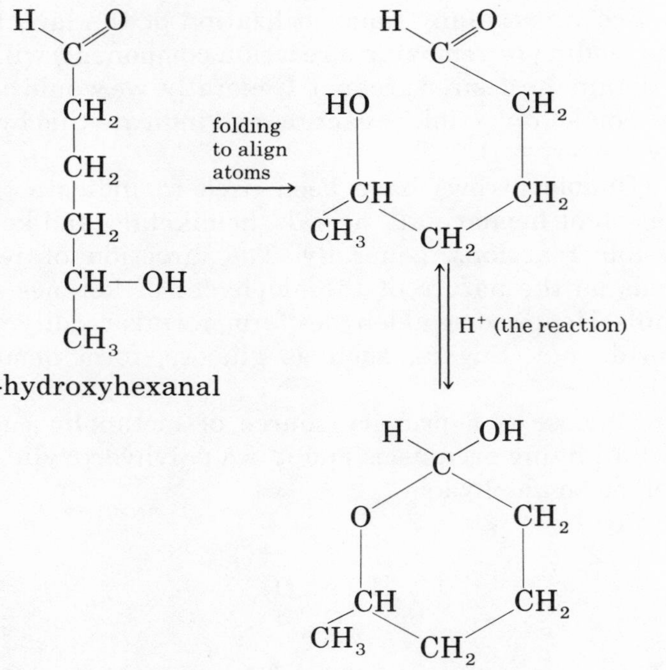

5-hydroxyhexanal

folding to align atoms →

H^+ (the reaction)

hemiacetal of 5-hydroxyhexanal

Recall that the bond angle between carbon atoms is 109.5°, not 180° as shown in the straight-chain form of 5-hydroxyhexanal. This geometry permits the tail of the molecules to "curl" around so that the hydroxyl group and aldehyde group come close together, which is necessary for the reaction to occur.

The reaction for glucose is identical except that four other hydroxyl groups, which do not participate in the reaction, are present:

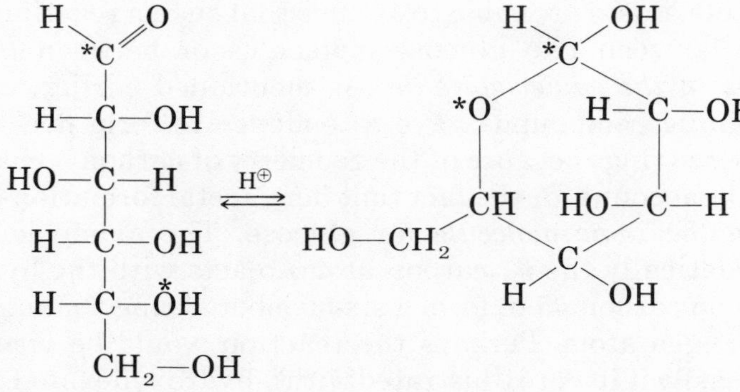

$H^{\oplus}$ ⇌

glucose
(straight-chain form)

glucose
(hemiacetal form)

The hemiacetal form of glucose usually is drawn by rotating the structure 90°. The carbon atoms within the ring are not shown; instead the vertexes of the ring structure stand for carbon atoms. Note that the *oxygen member* of the ring *is shown* with the appropriate chemical symbol:

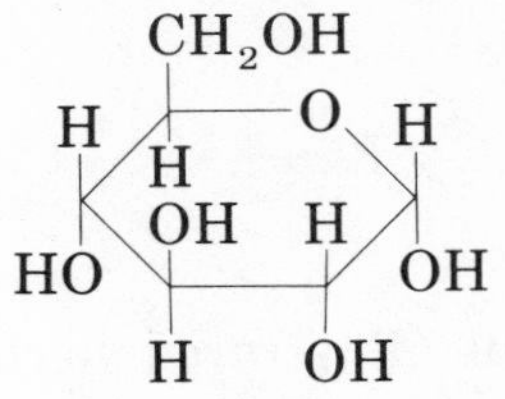

```
           CH2OH
            |
     H      /------O   H
       \  / H        \ /
        |              |
        \  OH    H    /
   HO    \         /    OH
            |-------|
            H       OH
```

hemiacetal form of glucose

Aldehydes, but not ketones, are oxidized easily to *carboxylic acids*. The aldehyde group may be oxidized by metal ions such as Cu^{++}:

$$R-\overset{\overset{\displaystyle O}{\|}}{C}-H + 2\,Cu^{++} + 4\,(OH^-) \longrightarrow$$

an aldehyde

$$R-\overset{\overset{\displaystyle O}{\|}}{C}-OH + \underset{(ppt)}{Cu_2O} + 2\,H_2O$$

a carboxylic acid

This reaction is the basis of Fehling's or Benedict's test for reducing sugars, such as glucose.

3.6 RESONANCE AND TAUTOMERIZATION

Because of the strong electronegativity of oxygen, carbonyl groups ($-\overset{\overset{\displaystyle O}{\|}}{C}-$) are polar. The Lewis structure of a ketone is given below. The x's and dots represent the electrons of oxygen and carbon atoms respectively. Imagine the situation where the oxygen and carbon atoms continue to share one pair of electrons but the oxygen assumes complete command of the other pair represented by the other half of the double bond:

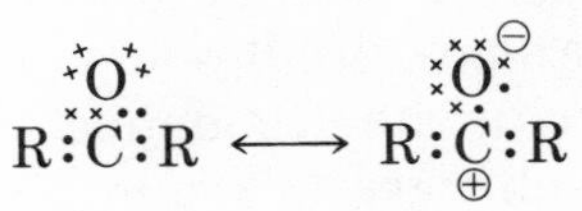

$$R:\overset{\overset{\displaystyle \ddot{O}}{\scriptsize \times\times\cdot\cdot}}{C}:R \longleftrightarrow R:\overset{\overset{\displaystyle \ddot{O}^{\ominus}}{\scriptsize \times\cdot}}{\underset{\oplus}{C}}:R$$

With such a shift the oxygen atom has gained an electron, giving it a negative charge, and the carbon atom has lost one electron, giving it a positive charge. Using bonds, the same electron shifts are depicted as

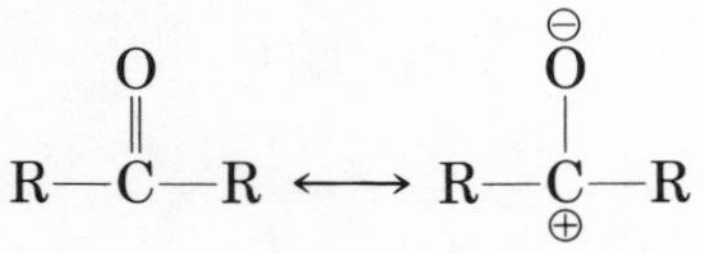

$$\mathrm{R{-}\overset{\overset{\displaystyle O}{\|}}{C}{-}R} \longleftrightarrow \mathrm{R{-}\underset{\oplus}{\overset{\overset{\displaystyle \overset{\ominus}{O}}{|}}{C}}{-}R}$$

The shift of electrons illustrated here is called *resonance*. *Resonance structures* are structures that have *identical empirical formulas* and the same arrangement of atoms, but the *bonding between certain atoms is different*. In the preceding examples the structural form on the left shows two covalent bonds between the carbon and oxygen atoms. The structural form on the right shows one covalent bond and one ionic bond ($\ominus$ and $\oplus$) between the carbon and oxygen atoms. The two forms of the ketone are called *resonance forms* and represent the extremes of molecular structures. The usual form is a hybrid or a compromise structure of the two extremes. This *does not* mean that the hybrid form must be an average of the two forms. In the case of ketones the characteristics of the hybrid form depend on the presence of other functional groups in the molecule, but generally the hydrid form is more like the keto form ($\mathrm{R{-}\overset{\overset{\displaystyle O}{\|}}{C}{-}R}$). However, having only partial characteristics of the ionic form means that there should be a slight unequal distribution of the electrons between the oxygen and carbon atoms. Thus perhaps the most accurate representation of the hybrid structure is

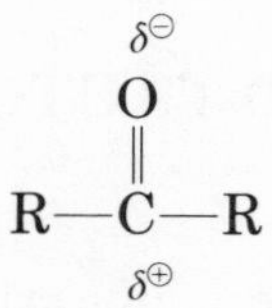

$$\mathrm{R{-}\underset{\delta^{\oplus}}{\overset{\overset{\displaystyle \overset{\delta^{\ominus}}{O}}{\|}}{C}}{-}R}$$

For the sake of simplicity the partial charges ($\delta^{\oplus}$ and $\delta^{\ominus}$) are usually omitted. However, the polarity of the carbonyl group ($\mathrm{{-}\overset{\overset{\displaystyle O}{\|}}{C}{-}}$) is due to its hybrid form.

Resonance is indicated by a double-headed arrow to distinguish it from a reversible chemical reaction, which is indicated by two single-headed arrows pointing in opposite directions. Aldehydes also exhibit resonance and polarity. You should be able to write the resonance forms and show the locations of partial charges ($\delta^{\oplus}$ and $\delta^{\ominus}$) for aldehydes by replacing one of the R-groups of the ketone with a hydrogen atom. Because of the unshared electrons of the oxygen atom, along with the oxygen atom's partial negative charge, which may attract positively charged hydrogen atoms, hydrogen bonding of ketones and aldehydes to water is possible. Because of the polarity of aldehyde and ketone groups, their presence increases the solubility of an organic compound in water:

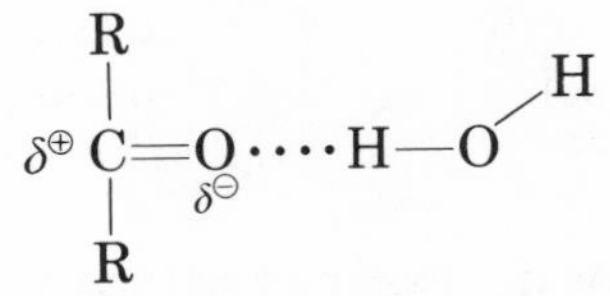

$$\overset{\delta\oplus}{} \underset{R}{\overset{R}{|}}C{=}\underset{\delta\ominus}{O}\cdots H{-}O{-}H$$

Hydrogen bonding of an aldehyde or ketone and water

Tautomerization is somewhat similar to resonance except that the position of a hydrogen *atom is changed* within a molecule. The two *tautomers* of acetone are

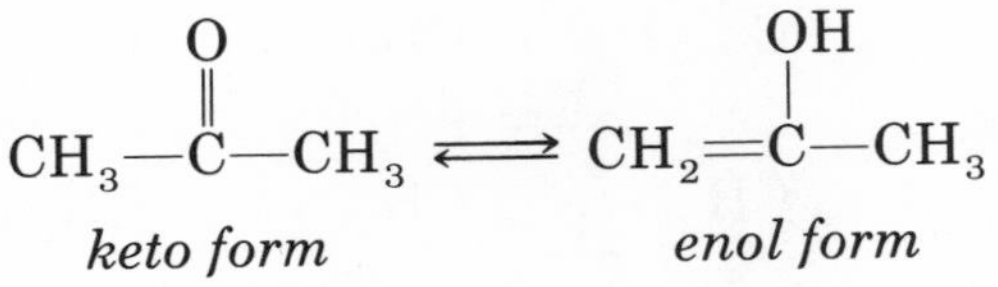

$$\underset{\textit{keto form}}{CH_3{-}\overset{O}{\overset{\|}{C}}{-}CH_3} \rightleftharpoons \underset{\textit{enol form}}{CH_2{=}\overset{OH}{\overset{|}{C}}{-}CH_3}$$

The keto form is the usual form for ketones. The other tautomer has a double bond between two carbon atoms, which is indicated by the *en-*, and a hydroxy group, which is indicated by *-ol*. Thus the term *enol* is descriptive of the structure although the strict rules for naming compounds are not followed.

Tautomerization is shown as a reversible reaction (although in this case the reaction lies to the left), and it is a reaction, not resonance. To change from the keto form to the enol form one of the hydrogen atoms of the methyl groups must come off the molecule and one hydrogen atom must be added to the oxygen atom to form the hydroxyl group. It is convenient to imagine that one hydrogen atom moves from one position to another in the reaction with rearrangements of electrons. In a strict sense this is not true although the net result is the same.

Pyruvic acid is an important intermediate in reactions used by living organisms to generate energy. It has a keto group ($-\overset{O}{\overset{\|}{C}}-$), a methyl group ($CH_3-$), and a carboxylic acid group ($-\overset{O}{\overset{\|}{C}}-OH$). Carboxylic acids will be discussed in the next chapter. For now, the carboxylic acid group will be ignored. However the methyl and keto groups can undergo the same tautomerization reaction illustrated for acetone. Thus pyruvic acid can exist in a keto form and an enol form:

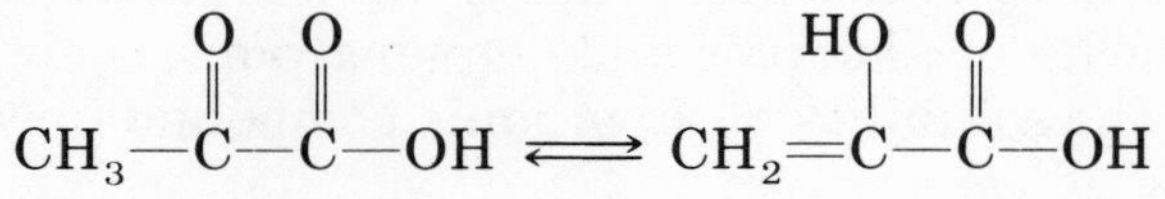

$$\underset{\substack{\text{pyruvic acid}\\ (\textit{keto form})}}{CH_3-\overset{\overset{\displaystyle O}{\|}}{C}-\overset{\overset{\displaystyle O}{\|}}{C}-OH} \rightleftharpoons \underset{\substack{\text{pyruvic acid}\\ (\textit{enol form})}}{CH_2=\overset{\overset{\displaystyle HO}{|}}{C}-\overset{\overset{\displaystyle O}{\|}}{C}-OH}$$

A methyl group is not necessary for tautomerization to occur. A methylene group ($-CH_2-$) next to a keto group ($-\overset{\overset{\displaystyle O}{\|}}{C}-$) will react in the same manner, as illustrated here for cyclohexanone:

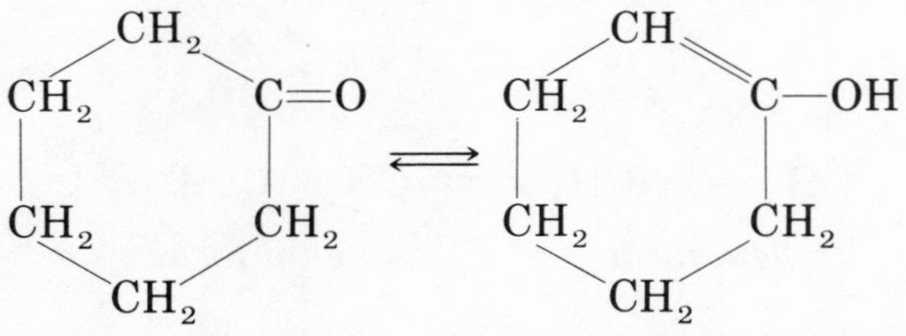

$$\text{ring: } CH_2, CH_2, CH_2, CH_2, CH_2, C{=}O \;\rightleftharpoons\; \text{ring: } CH{=}C{-}OH, CH_2, CH_2, CH_2, CH_2$$

cyclohenanone (*keto form*) ⇌ cyclohexanone (*enol form*)

EXERCISES

1. Draw the Lewis structure for ethanol.
2. Name the following compounds.
   (a) $CH_3-CH_2-\underset{\underset{\displaystyle OH}{|}}{CH}-CH_2-CH_2-CH_3$
   (b) $CH_3-CH_2-CH_2-\underset{\underset{\displaystyle OH}{|}}{CH}-CH_2-CH_3$
   (c) $CH_3-\underset{\underset{\displaystyle OH}{|}}{CH}-CH=CH_2$
   (d) $CH_3-CH_2-CH_2-CH_2-CH_2-CH_2-CH_2-CH_2-OH$
3. Draw the structures for the following compounds.
   (a) 2,3-butandiol
   (b) 3-methyl-2-hexanol
   (c) 4-ethyl-2-hexanol
   (d) 1,3-propandiol

4. Draw all of the structures for the skeletal and positional isomers of the alcohols with a molecular formula $C_4H_{10}O$ (C_4H_9-OH). (*Hint:* There are four isomers.)

5. Indicate whether each of the following structures is a primary, secondary, or tertiary alcohol.

(a) CH_3-CH_2-OH

(b) $$\begin{array}{ccccccc} & & CH_3 & & & & \\ & & | & & & & \\ CH_3 & - & C & - & CH_2 & - & CH_3 \\ & & | & & & & \\ & & OH & & & & \end{array}$$

(c) $$\begin{array}{ccccccc} CH_3 & - & CH & - & CH_2 & - & CH_3 \\ & & | & & & & \\ & & OH & & & & \end{array}$$

(d) $$\begin{array}{ccccccc} CH_3 & - & CH_2 & - & CH_2 & - & CH_2 \\ & & & & & & | \\ & & & & & & OH \end{array}$$

6. Draw the Lewis structure for formaldehyde.

7. Name the following compounds.

(a) $$\begin{array}{ccccccccc} & & & & & & O & & \\ & & & & & & \| & & \\ CH_3 & - & CH_2 & - & CH_2 & - & C & - & H \end{array}$$

(b) $$\begin{array}{ccccccccccc} & & & & & & O & & & & \\ & & & & & & \| & & & & \\ CH_3 & - & CH_2 & - & CH_2 & - & C & - & CH_2 & - & CH_3 \end{array}$$

(c) $$\begin{array}{ccccc} & & O & & \\ & & \| & & \\ CH_3 & - & C & - & CH_3 \end{array}$$

(d) $$\begin{array}{ccccccccccccc} & & O & & & & & & O & & & & \\ & & \| & & & & & & | & & & & \\ CH_3 & - & C & - & CH_2 & - & CH_2 & - & C & - & CH_2 & - & CH_3 \end{array}$$

8. Complete the following reactions by drawing the structures for the missing components.

(a) $$\xrightarrow{\text{mild oxidation}} \begin{array}{ccccccc} & & & & O & & \\ & & & & \| & & \\ CH_3 & - & CH_2 & - & C & - & H \end{array}$$

(b) $$\begin{array}{ccccccc} & & & & OH & & \\ & & & & | & & \\ CH_3 & - & CH_2 & - & CH & - & CH_3 \end{array} + [O] \longrightarrow \qquad\qquad + HOH$$

9. Complete the following reactions.

(a) $$\begin{array}{ccccc} & & O & & \\ & & \| & & \\ CH_3 & - & C & - & H \end{array} + CH_3-CH_2-OH \overset{H^{\oplus}}{\rightleftharpoons}$$

(b) $$\begin{array}{ccccc} & & OH & & \\ & & | & & \\ CH_3 & - & C & - & H \\ & & | & & \\ & & O & - & CH_2-CH_3 \end{array} + CH_3-CH_2-OH \overset{H^{\oplus}}{\rightleftharpoons}$$

10. Complete the following reactions.

(a) $CH_3—\overset{\overset{\displaystyle O}{\|}}{C}—CH_2—CH_3 + H_2 \xrightarrow{\text{catalyst}}$

(b) $CH_3—\overset{\overset{\displaystyle O}{\|}}{C}—H + 2\,Cu^{++} + 4\,(OH^-) \longrightarrow$

Carboxylic Acids

The general formula for a carboxylic acid is

$$R{-}\overset{\overset{\Large O}{\|}}{C}{-}OH \qquad R{-}COOH \qquad \text{or} \qquad R{-}CO_2H$$

and the carboxyl group or carboxylic acid group is written as

$$-\overset{\overset{\Large O}{\|}}{C}{-}OH \qquad -COOH \qquad \text{or} \qquad -CO_2H$$

When —COOH or —CO_2H is written, it should be understood that one oxygen atom is double bonded to the carbon and the other oxygen atom is a part of an —OH-group.

Earlier, oxidation was defined as the addition of oxygen and/or removal of hydrogen whereas reduction was defined as the addition of hydrogen and/or removal of oxygen. Hence the addition of bonds between a carbon atom and oxygen atom(s) or loss of bonds between a carbon atom and hydrogen atom(s) indicates oxidation of the affected carbon atom. Consider the oxidation state of carbon in the following series:

$$H{-}\overset{\overset{\large H}{|}}{\underset{\underset{\large H}{|}}{C}}{-}H \qquad \text{methane}$$

$$\mathrm{H{-}\underset{\underset{H}{|}}{\overset{\overset{H}{|}}{C}}{-}O{-}H} \qquad \text{methyl alcohol}$$

$$\mathrm{H{-}\overset{\overset{O}{\|}}{C}{-}H} \qquad \text{formaldehyde}$$

$$\mathrm{H{-}\overset{\overset{O}{\|}}{C}{-}O{-}H} \qquad \text{formic acid}$$

$$\mathrm{O{=}C{=}O} \qquad \text{carbon dioxide}$$

For each of these compounds, count the number of bonds between the carbon atom and oxygen atom(s). There are 0, 1, 2, 3, and 4 as you proceed down the series. Note that ketones have two bonds between carbon and oxygen; hence they have the same oxidation state as aldehydes. A carboxylic acid group represents the highest oxidation state of carbon in an organic compound. If it seems as if the point has been overemphasized, just keep in mind that, biochemically, animals get their energy from the oxidation of carbon compounds by atmospheric O_2. Of course, CO_2 and H_2O are the end products. Incidentally, we could count the number of bonds between C and H to see that the most reduced form (methane) has the highest number of bonds. This supports the notion that reduction represents the reverse of oxidation.

Formic acid, $\mathrm{H{-}\overset{\overset{O}{\|}}{C}{-}OH}$, is the simplest organic acid. Acetic acid, $CH_3{-}COOH$, is next. Butyric acid is a four-carbon acid:

$$\mathrm{CH_3{-}CH_2{-}CH_2{-}\overset{\overset{O}{\|}}{C}{-}OH}$$

butyric acid (trivial)
butanoic acid (Geneva)

In the Geneva system the corresponding alkane's name is changed by dropping the *-e* and adding an *-oic* and the word *acid*. Hence *butane* becomes *butanoic acid*. And what about numbers? Generally they are not needed because an acid group can occur only at the end of a chain; in fact, the carbon atom of the carboxyl group automatically becomes the *#1 carbon* atom. This is illustrated in the following compound:

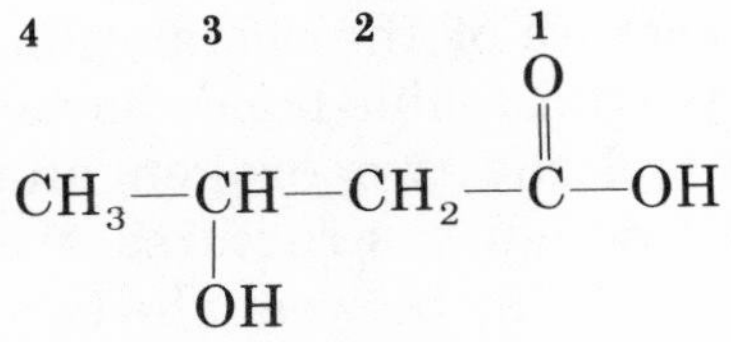

$$\overset{4}{CH_3}-\underset{\underset{OH}{|}}{\overset{3}{CH}}-\overset{2}{CH_2}-\overset{\overset{1}{O}}{\overset{\|}{C}}-OH$$

3-hydroxybutanoic acid

Since these compounds are called acids, they must donate protons ($H^{\oplus}$) in aqueous solutions. Which hydrogen atom ionizes? The hydrogen attached to the oxygen of the carboxylic acid group does ionize easily, unlike the —OH-group in alcohols. This is due to an extra oxygen atom attached to the same carbon atom:

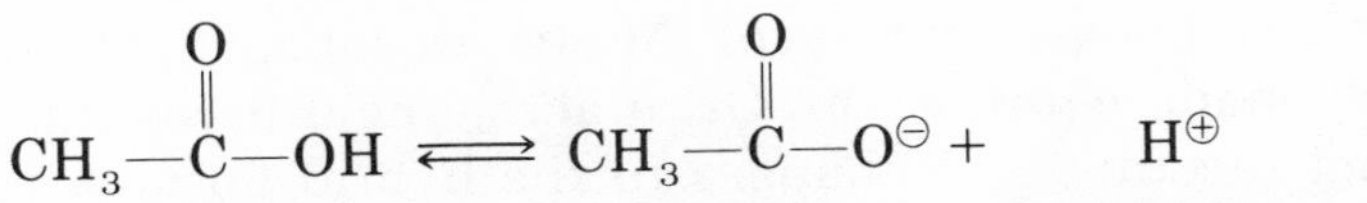

$$CH_3-\overset{\overset{O}{\|}}{C}-OH \rightleftharpoons CH_3-\overset{\overset{O}{\|}}{C}-O^{\ominus} + H^{\oplus}$$

acetic acid — acetate ion — hydrogen ion

ionization of acetic acid

The nature of the acetate ion explains why dissociation, and hence formation of hydrogen ions, is favored. The resonance forms of the acetate ion are involved in the explanation:

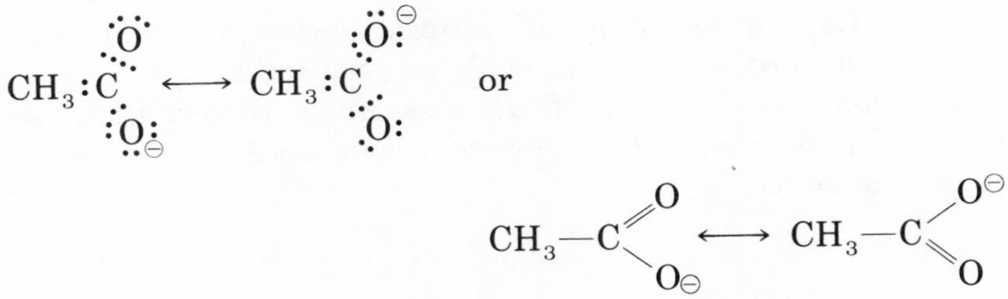

$$CH_3:C\begin{matrix}\ddot{O}\\ \\ :\ddot{O}:^{\ominus}\end{matrix} \longleftrightarrow CH_3:C\begin{matrix}:\ddot{O}:^{\ominus}\\ \\ \ddot{O}:\end{matrix}$$

or

$$CH_3-C\begin{matrix}\nearrow\!\!\!\!= O\\ \searrow O^{\ominus}\end{matrix} \longleftrightarrow CH_3-C\begin{matrix}\nearrow O^{\ominus}\\ \searrow\!\!\!\!= O\end{matrix}$$

resonance structures of the acetate ion

Once ionized, the acetate ion could shift the extra electron (from the H atom) and one pair of electrons in the double bond between the two oxygen atoms. We cannot distinguish between the two structures just given, but *on the average*, each oxygen could carry only a $\frac{1}{2}$ negative charge rather than a whole negative charge:

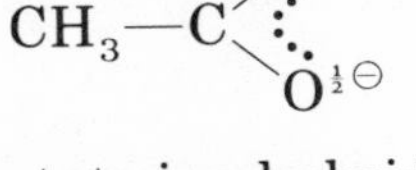

$$CH_3-C\begin{matrix}\nearrow O^{\frac{1}{2}\ominus}\\ \vdots\\ \searrow O^{\frac{1}{2}\ominus}\end{matrix}$$

acetate ion hybrid

The real structure of the acetate ion is a hybrid with partial single and partial double-bond characteristics between the carbon atom and the two oxygen atoms. Unlike with ketones, the two resonance structures are equivalent and hence the hybrid is halfway between the two extremes. Recall that resonance structures have identical molecular formulas but the bonding between certain atoms is different. The possibility of resonance (the shifting back and forth of bonds or electrons between different atoms in the same molecule) or the existence of a hybrid or intermediate form in a compound makes it more stable than if resonance could not occur.

How would resonance make acetic acid a stronger acid than it would be if resonance did not occur? Since resonance does not occur in the undissociated form, resonance should not affect the dissociation step. However, we visualize the dissociation and reassociation of the acetate ion and proton to be in a dynamic equilibrium. Once acetic acid dissociates, the product (acetate ion) changes to the hybrid form. A proton would not be attracted to one of the oxygen atoms of the hybrid acetate ion nearly as strongly (only $\frac{1}{2}$ negative charge) as it would be attracted to an oxygen atom bearing a full negative charge. Thus the formation of a hybrid interferes with the reassociation step in the equilibrium. The net result is a shift in the equilibrium toward dissociation, which means that the hydrogen-ion concentration should be greater, and the strength of an acid is directly proportional to hydrogen-ion concentration.

Although resonance of ionized carboxylic acids was illustrated here with acetic acid, we can see that *resonance should and does occur with all dissociated carboxylic acids because the R-group (CH_3— for acetic acid) does not participate in the resonance.*

4.1 PROPERTIES OF CARBOXYLIC ACIDS

Carboxylic acid groups are more polar than alcohols, aldehydes, or ketones. The fact that they can ionize greatly increases their solubility in water. The pK_a's of carboxylic acids are generally around 5.0.

Let us review the meaning of pK_a using as an example acetic acid:

$$CH_3—\overset{\overset{\displaystyle O}{\|}}{C}—OH \rightleftharpoons CH_3—\overset{\overset{\displaystyle O}{\|}}{C}—O^{\ominus} + H^{\oplus}$$

$$K_a = \frac{[\mathrm{CH_3{-}\overset{\overset{\displaystyle O}{\|}}{C}{-}O^{\ominus}}][\mathrm{H^{\oplus}}]}{[\mathrm{CH_3{-}\overset{\overset{\displaystyle O}{\|}}{C}{-}OH}]}$$

$$pK_a = -\log K_a$$

(Remember the convention of using brackets to mean concentration, generally in moles per liter)

The pK_a of an acid is simply the negative of the logarithm (to the base 10) of the equilibrium constant for the ionization reaction. In the case of acetic acid the equilibrium constant (sometimes called a dissociation constant) is 1.76×10^{-5}; that is,

$$K_a = 1.76 \times 10^{-5}$$

Therefore

$$pK_a = -\log K_a = (-1)\log K_a = (-1)\log(1.76 \times 10^{-5})$$
$$= (-1)(-4.75) = 4.75$$

Let $\quad \mathrm{HA} = \mathrm{CH_3{-}\overset{\overset{\displaystyle O}{\|}}{C}{-}OH}$ and $\mathrm{A^{\ominus}} = \mathrm{CH_3{-}\overset{\overset{\displaystyle O}{\|}}{C}{-}O^{\ominus}}$

Then $\quad K_a = \dfrac{[\mathrm{A^-}][\mathrm{H^{\oplus}}]}{[\mathrm{HA}]} = \dfrac{[\mathrm{A^{\ominus}}]}{[\mathrm{HA}]} \cdot \dfrac{[\mathrm{H^{\oplus}}]}{1}$

Taking the logarithms of both sides of the equation and transposing terms gives

$$\log K_a = \log \frac{[\mathrm{A^{\ominus}}]}{[\mathrm{HA}]} + \log [\mathrm{H^{\oplus}}]$$

$$-\log [\mathrm{H^{\oplus}}] + \log K_a = \log \frac{[\mathrm{A^{\ominus}}]}{[\mathrm{HA}]}$$

$$-\log \mathrm{H^{\oplus}} = -\log K_a + \log \frac{[\mathrm{A^{\ominus}}]}{[\mathrm{HA}]}$$

Since $-\log [\mathrm{H^{\oplus}}] = \mathrm{pH}$ and $-\log K_a = pK_a$, then

$$\mathrm{pH} = pK_a + \log \frac{[\mathrm{A^{\ominus}}]}{[\mathrm{HA}]}$$

When $[\mathrm{A^{\ominus}}] = [\mathrm{HA}]$, $[\mathrm{A^{\ominus}}]/[\mathrm{HA}] = 1$. Because $\log 1 = 0$,

$$\mathrm{pH} = pK_a + \log 1 = pK_a + 0$$

or $\mathrm{pH} = pK_a$ when $[\mathrm{A^{\ominus}}] = [\mathrm{HA}]$. This means that the concentration of the ionized acid, $[\mathrm{A^-}]$, is equal to the concentration of the un-ionized form, [HA], when the pH of the *solution* is 4.75. Another way to define the pK_a is to say that it is numerically equal to the pH of the solution when the acid is one-half ionized.

If enough NaOH were added to bring the pH of a solution of acetic acid up to 7, what would be the relative concentrations of $A^{\ominus}$ and HA? The equation just derived can be used:

$$\text{pH} = 7, pK_a = 4.75, \log \frac{[A^{\ominus}]}{[HA]} = ?$$

$$\text{pH} = pK_a + \log \frac{[A^{\ominus}]}{[HA]}$$

$$7 = 4.75 + \log \frac{[A^{\ominus}]}{[HA]}$$

$$2.25 = \log \frac{[A^{\ominus}]}{[HA]}$$

$$178 = \frac{[A^{\ominus}]}{[HA]}$$

This specifies that there are 178 acetate ions ($A^{\ominus}$) for each un-ionized acetic acid molecule (HA) at pH 7.

Consider a general representation of the ionization of a carboxylic acid, where

$$HA = R{-}\overset{\overset{\displaystyle O}{\|}}{C}{-}OH \qquad \text{and} \qquad A^{\ominus} = R{-}\overset{\overset{\displaystyle O}{\|}}{C}{-}O^{\ominus}$$

$$HA \rightleftharpoons A^{\ominus} + H^{\oplus}$$

If K_a = the equilibrium constant for the ionization reaction, $pK_a = -\log K_a$, $\text{pH} = -\log [H^{\oplus}]$, then an equation identical to the one derived for acetic acid may be written:

$$\text{pH} = pK_a + \log \frac{[A^{\ominus}]}{[HA]}$$

This general formula is called the Henderson-Hasselbalch equation, and it applies to acids in general.

Compared to the common inorganic acids such as HCl, HNO_3, H_2SO_4, and H_3PO_4, carboxylic acids are fairly weak. You will recall from inorganic chemistry that in a 1 M solution, HCl is considered to be totally ionized; that is, $[H^{\oplus}] = 1$ M and $[Cl^{\ominus}] = 1$ M. In contrast, a 1 M solution of acetic acid contains about 0.99581 M $CH_3{-}\overset{\overset{\displaystyle O}{\|}}{C}{-}OH$, about 0.00419 M $CH_3{-}\overset{\overset{\displaystyle O}{\|}}{C}{-}O^{\ominus}$, and about

0.00419 M $H^{\oplus}$, or pH = 2.38. Vinegar contains 5% acetic acid (about 0.83 M) and has a pH of about 2.4.

4.2 SALTS OF CARBOXYLIC ACIDS

Since inorganic acids react with inorganic bases, it should be expected that carboxylic acids react with inorganic bases, such as KOH and NaOH, as well:

$$\underset{\text{acetic acid}}{CH_3\overset{\overset{\displaystyle O}{\|}}{C}{-}OH} + NaOH \longrightarrow \underset{\substack{\text{sodium acetate}\\ \textit{(an organic salt)}}}{CH_3{-}\overset{\overset{\displaystyle O}{\|}}{C}{-}O^{\ominus}Na^{\oplus}} + H_2O$$

The reaction is called neutralization. The general neutralization reaction is

$$R{-}\overset{\overset{\displaystyle O}{\|}}{C}{-}OH + NaOH \longrightarrow R{-}\overset{\overset{\displaystyle O}{\|}}{C}{-}O^{\ominus}Na^{\oplus} + H_2O$$

In naming the salts or ions of acids, the acid's name is changed by dropping the *-ic*, adding *-ate*, and omitting *acid*. Hence *butanoic acid* becomes the *butanoate ion* when ionized, and the name *butyric acid* becomes *butyrate*; likewise a *carboxylic acid* group becomes a *carboxylate* group when it ionizes. This rule holds for trivial and Geneva names.

The salts ($Na^{\oplus}$ or $K^{\oplus}$) of long-chain carboxylic acids are called soaps; two examples are sodium palmitate and sodium stearate:

$$CH_3{\leftarrow}CH_2{\rightarrow}_{14}\overset{\overset{\displaystyle O}{\|}}{C}{-}O^{\ominus}Na^{\oplus}$$

sodium palmitate (trivial)
sodium hexadecanoate (Geneva)

$$CH_3{\leftarrow}CH_2{\rightarrow}_{16}\overset{\overset{\displaystyle O}{\|}}{C}{-}O^{\ominus}Na^{\oplus}$$

sodium stearate (trivial)
sodium octadecanoate (Geneva)

Table 4.1 give the names and structures of some other biologically important carboxylic acids. Long-chain carboxylic acids are generally derived from fats such as lard and beef tallow. Because of their usual source, these long-chain acids are also called fatty acids.

Detergents are compounds that can emulsify oil and grease in water, and soap is one type of detergent. Commercial soaps generally are sodium salts of a *mixture* of fatty acids. Sodium stearate would be present in essentially all commercial soaps but it would not be an exclusive compound.

Different portions of a soap molecule have quite different physical properties. The long hydrocarbon chain is strongly hydrophobic whereas the carboxylate group is polar and quite hydrophilic. These portions of a soap molecule are shown diagrammatically in Fig. 4.1. When placed in water the salt ionizes, yielding a negatively charged polar head. The conflicting physical properties of the hydrophobic tail and the hydrophilic polar head prevent soap molecules from forming true solutions in which solvent molecules (water) surround

TABLE 4.1
Names and Structures for Some Important Carboxylic Acids

Trivial Name	Structure	Geneva Name	Melting Point (°C)
		Monocarboxylic	
Formic*	$H—CO_2H$	Methanoic acid	8.4
Acetic*	$CH_3—CO_2H$	Ethanoic acid	16.6
Propionic*	$CH_3—CH_2—CO_2H$	Propanoic acid	−20.8
Butyric*	$CH_3\text{(}CH_2\text{)}_2CO_2H$	Butanoic acid	− 4.3
Isobutric	$CH_3—CH(CH_3)—CO_2H$	2-Methylpropanoic acid	−46.1
Valeric*	$CH_3\text{(}CH_2\text{)}_3CO_2H$	Pentanoic acid	34
Caproic	$CH_3\text{(}CH_2\text{)}_4CO_2H$	Hexanoic acid	− 2
Caprylic	$CH_3\text{(}CH_2\text{)}_6CO_2H$	Octanoic acid	16.5
Capric	$CH_3\text{(}CH_2\text{)}_8CO_2H$	Decanoic acid	31.5
Lauric	$CH_3\text{(}CH_2\text{)}_{10}CO_2H$	Dodecanoic acid	44
Myristic	$CH_3\text{(}CH_2\text{)}_{12}CO_2H$	Tetradecanoic acid	58
Palmitic*	$CH_3\text{(}CH_2\text{)}_{14}CO_2H$	Hexadecanoic acid	63
Stearic*	$CH_3\text{(}CH_2\text{)}_{16}CO_2H$	Octadecanoic acid	72
		Dicarboxylic	
Oxalic*	$HO_2C—CO_2H$	Ethanedioic acid	
Malonic*	$HO_2C—CH_2—CO_2H$	Propanedioic acid	
Succinic*	$HO_2C\text{(}CH_2\text{)}_2CO_2H$	Butanedioic acid	
Glutaric*	$HO_2C\text{(}CH_2\text{)}_3CO_2H$	Pentanedioic acid	
Adipic	$HO_2C\text{(}CH_2\text{)}_4CO_2H$	Hexanedioic acid	
Pimelic	$HO_2C\text{(}CH_2\text{)}_5CO_2H$	Heptanedioic acid	
Suberic	$HO_2C\text{(}CH_2\text{)}_6CO_2H$	Octanedioic acid	
Azelaic	$HO_2C\text{(}CH_2\text{)}_7CO_2H$	Nonanedioic acid	
Sebacic	$HO_2C\text{(}CH_2\text{)}_8CO_2H$	Decanedioic acid	

* These acids are quite important in biochemistry.

individual solute (soap) molecules. Instead the hydrocarbon tails are attracted to one another; that is, they tend to "dissolve" in one another. This creates a small spherical structure called a *micelle* (see Fig. 4.2), in which the hydrocarbon tails are directed toward the interior of the sphere and the polar head toward the surface of the sphere. Micelle structures form spontaneously. The arrangement of molecules allows the polar heads to be near water molecules while the hydrocarbon tails are near other hydrocarbon tails and water is excluded from the interior to a large extent. Compared to many soluble compounds, micelles are large. Their size causes the aqueous solution to scatter light, which gives soap suspensions their milky appearance.

FIGURE 4.1

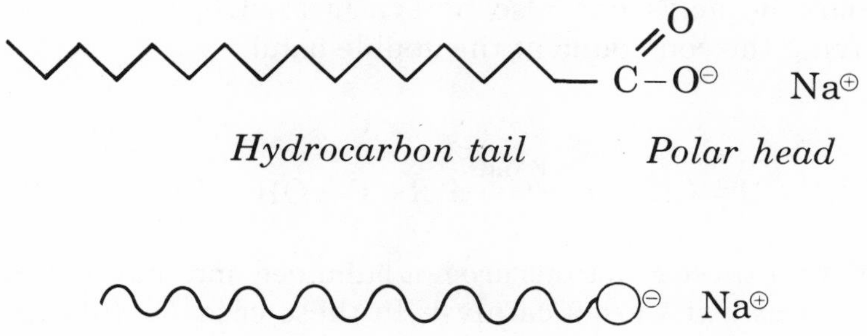

$C(=O)-O^{\ominus}$ $Na^{\oplus}$

Hydrocarbon tail *Polar head*

$Na^{\oplus}$

Representations of Soap Molecules. Top, the sawtooth structure of sodium palmitate; bottom, a general diagram for the sodium salt of any long-chain fatty acid, such as myristic, palmitic, or stearic acid.

FIGURE 4.2

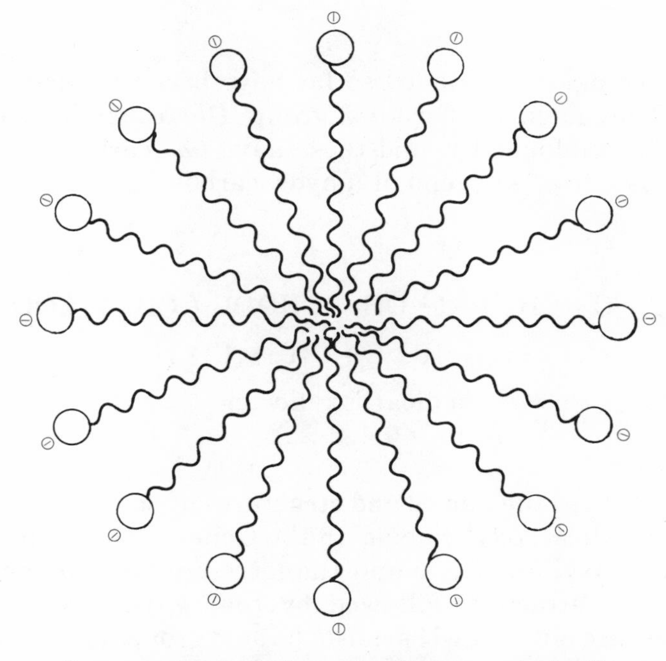

Cross-sectional view of a detergent micelle

4.3 SYNTHESIS OF CARBOXYLIC ACIDS

Oxidation of aldehydes and primary alcohols yields carboxylic acids. A strong oxidizing agent such as $KMnO_4$ or $K_2Cr_2O_7$ may be used:

$$R{-}CH_2{-}OH \xrightarrow{KMnO_4} R{-}\overset{\overset{\large O}{\|}}{C}{-}OH$$

$$R{-}\overset{\overset{\large O}{\|}}{C}{-}H \xrightarrow{KMnO_4} R{-}\overset{\overset{\large O}{\|}}{C}{-}OH$$

Carboxylic acids can also be synthesized from alkenes, by cleaving the compound at the double bond:

$$R{-}CH{=}CH{-}R' \xrightarrow{KMnO_4} R{-}\overset{\overset{\large O}{\|}}{C}{-}OH + HO{-}\overset{\overset{\large O}{\|}}{C}{-}R'$$

Note that these reactions are not balanced and that $KMnO_4$ is written as if it were a catalyst. In these cases $KMnO_4$ is the oxidizing agent and is given to indicate strong oxidizing conditions.

4.4 DICARBOXYLIC ACIDS

All of the acids mentioned so far have been *mono*carboxylic acids. They have *one* carboxyl group. *Di*carboxylic acids are important biologically, and these have *two* carboxyl groups, generally one at each end of a hydrocarbon chain:

$$HO{-}\overset{\overset{\large O}{\|}}{C}{+}CH_2{+}_n\overset{\overset{\large O}{\|}}{C}{-}OH \qquad HOOC{+}CH_2{+}_nCOOH$$

$$\text{or} \quad HO_2C{+}CH_2{+}_nCO_2H$$

a dicarboxylic acid
($n = 0, 1, 2, \ldots$)

Table 4.1 gives the names and structures of some biologically important dicarboxylic acids. In the Geneva system the name of a dicarboxylic acid is simply the corresponding alkane name with *-dioic* attached, followed by the separate word *acid.* Numbers are not needed because the acid groups must be at the ends of the chain.

Because of the general nature of reactions, it should be expected that both carboxyl groups would react similarly. Thus two molecules of NaOH should react with one molecule of a dicarboxylic acid.

EXERCISES

1. True or False.
   - **(a)** T F The oxidation state of the carbon atom in methanol is higher than that in methane.
   - **(b)** T F Formaldehyde is more reduced (less oxidized) than formic acid.
   - **(c)** T F Methane represents the highest oxidation state for carbon.
   - **(d)** T F Carbon dioxide is less oxidized than formic acid.
2. Name the following compounds.
   - **(a)** $CH_3-CH_2-CH_2-\overset{\overset{\large O}{\|}}{C}-OH$
   - **(b)** $HO-\overset{\overset{\large O}{\|}}{C}-CH_2-CH_2-\overset{\overset{\large O}{\|}}{C}-OH$
   - **(c)** $CH_3-\underset{\underset{\large CH_3}{|}}{CH}-CH_2-CH_2-\overset{\overset{\large O}{\|}}{C}-OH$
   - **(d)** $CH_3(CH_2)_{16}\overset{\overset{\large O}{\|}}{C}-OH$
3. Draw the Lewis structure for acetic acid.
4. Complete the following reactions.
   - **(a)** $R-\overset{\overset{\large O}{\|}}{C}-OH + NaOH \longrightarrow$
   - **(b)** $R-CH_2-OH \xrightarrow{KMnO_4}$
   - **(c)** $R-\overset{\overset{\large O}{\|}}{C}-H \xrightarrow{KMnO_4}$
5. Write the reaction for the ionization of propanoic acid.
6. The K_a for the ionization of propanoic acid is 1.343×10^{-5}. What is the pK_a for propanoic acid?
7. If in an aqueous solution the concentration of the ionized form of propanoic acid is exactly equal to the concentration of the un-ionized form, what is the pH? Information from Question 6 may be used.

8. If the pH is 8, what is the ratio of the concentrations of the ionized form to the un-ionized form, $[A^{\ominus}]/[HA]$, in a solution of propanoic acid?

9. If the concentration of the ionized form of propanoic acid, $[A^{\ominus}]$, is 0.001 M and the concentration of the un-ionized form, [HA], is 0.1 M, what is the pH of the solution?

10. If the concentration of the ionized form of propanoic acid, $[A^{\ominus}]$, is 0.1 M and the concentration of the un-ionized form, [HA], is 0.01 M, what is the pH of the solution?

Esters and Ethers

5

5.1 ESTERS

A very important reaction of acids is called esterification. This is a general reaction for acids and alcohols but is illustrated here using specific compounds. An inorganic acid such as HCl or H_2SO_4 will supply hydrogen ions (indicated by the $H^{\oplus}$ above the arrow), which catalyze the reaction:

$$\underset{\text{acetic acid}}{CH_3-\overset{\overset{\large O}{\|}}{C}-OH} + \underset{\text{ethyl alcohol}}{HO-CH_2-CH_3} \xrightleftharpoons{H^{\oplus}}$$

$$\underset{\text{ethyl acetate}}{CH_3-\overset{\overset{\large O}{\|}}{C}-O-CH_2-CH_3} + H_2O$$

The organic product of the reaction is ethyl acetate, which is an ester.

The general formula for an ester is

$$R-\overset{\overset{\large O}{\|}}{C}-O-R' \qquad R-\overset{\overset{\large O}{\|}}{C}OR' \qquad \text{or} \qquad RCOOR'$$

The R-groups in any one molecule may be the same alkyl radical or different.

In naming esters it is important to realize that a portion of the molecule looks like the carboxylate ion without a negative charge:

$$R—\overset{\overset{\displaystyle O}{\|}}{C}—O$$

The remainder of the ester molecule is simply an alkyl radical. Using the ester shown in the first reaction, the name is: *first*, the name of the alkyl radical corresponding to the alcohol moiety, *ethyl*, and, *second*, the name of the carboxylate ion or salt, *acetate*, which gives *ethyl acetate*. Note that the name for the alcohol radical is not joined to the acid's name. They are separate words. Many different esters of acetic acid are formed by using different alcohols. For example,

$$CH_3—\overset{\overset{\displaystyle O}{\|}}{C}—O—CH_3 \qquad CH_3—\overset{\overset{\displaystyle O}{\|}}{C}—O—CH_2—CH_2—CH_3$$

methyl acetate — *n*-propyl acetate

Different acids may be esterified with a particular alcohol:

$$H\overset{\overset{\displaystyle O}{\|}}{C}—O—CH_2—CH_3 \qquad CH_3CH_2—CH_2—\overset{\overset{\displaystyle O}{\|}}{C}—O—CH_2—CH_3$$

ethyl formate — ethyl butanoate

Reminder: In naming esters, remember that *an ester is a combination of a carboxylic acid and an alcohol.*

Esters may be named in the trivial system by using the word *ester*; for example, methyl acetate may be called "the methyl ester of acetic acid."

In organic chemistry the term *derivative* appears frequently. A derivative is simply a compound derived from another compound or other compounds. Hence the ester ethyl acetate is a derivative of acetic acid; ethyl acetate is also a derivative of ethanol.

Up to this point R-groups have been defined as alkyl radicals, but now that definition needs to be broadened. An R-group may represent an alkyl radical modified by the presence of other functional groups. The following are some simple examples:

$$—CH_2—CH{=}CH_2 \qquad —CH_2—OH$$

$$CH_2—CH_2—\overset{\overset{\displaystyle O}{\|}}{C}—CH_3 \qquad CH_2—\overset{\overset{\displaystyle O}{\|}}{C}—OH$$

Consider the compound

$$CH_3-\overset{\overset{\Large O}{\|}}{C}-O-CH_2-CH_2-CH=CH_2$$

The R-group of the alcohol portion of the molecule (1-hydroxy-3-butene) is not a simple word, and so using the conventional method of naming this ester is cumbersome. "The acetate derivative of 1-hydroxy-3-butene" describes the molecule completely and is an acceptable name.

5.1.1 Acyl Groups

The general formula

$$R-\overset{\overset{\Large O}{\|}}{C}-$$

represents an acyl group, which is a carboxylic acid with its —OH missing. Acyl groups do not exist alone, but are present in various compounds. An ester contains an acyl group, which frequently may be used in naming the derivative. To name an acyl group, the corresponding acid's name is modified by substituting the suffix *-yl* for the *-ic* and dropping the word *acid*. The acyl group corresponding to acetic acid is named acetyl, pronounced "a′·see·t'l" or "ass·a·teal′":

$$CH_3-\overset{\overset{\Large O}{\|}}{C}-$$

acetyl group

The same naming methods are followed for both trivial and Geneva names of acids:

$$CH_3\text{-}(CH_2)_3\overset{\overset{\Large O}{\|}}{C}-$$

valeryl or pentanoyl group

Ethyl formate and ethyl butanoate, two esters given previously, contain a formyl group and a butanoyl group, respectively.

For naming compounds, the acyl name is never used alone; instead it modifies the name of another compound. *Propanoyl chloride* describes the compound in which a propanoyl group has replaced the hydrogen atom of HCl, hydrogen chloride:

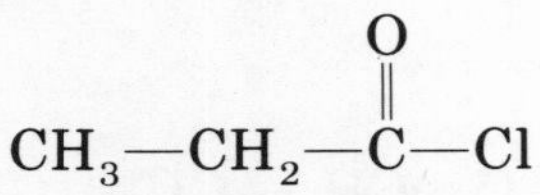

$$CH_3{-}CH_2{-}\overset{\overset{\large O}{\|}}{C}{-}Cl$$

propanoyl chloride

The name of the acyl group is given as the first (and separate) word, which is followed by the name of the molecule or atom modified.

Choline is an important biological compound. It has other functional groups, but it also contains a hydroxyl group (—OH). *Acetyl choline* is a name that describes the derivative of choline in which the hydrogen atom of the hydroxyl group has been replaced by an acetyl group:

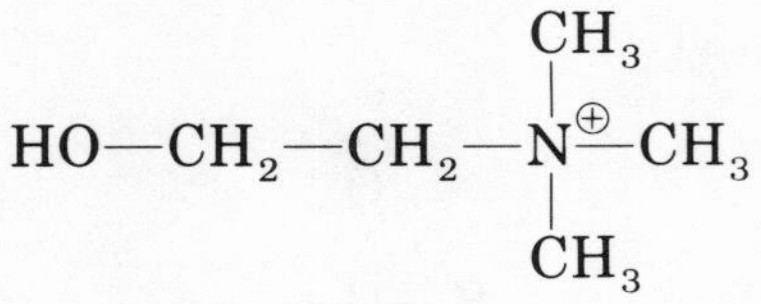

$$HO{-}CH_2{-}CH_2{-}\overset{\overset{\large CH_3}{|}}{\underset{\underset{\large CH_3}{|}}{N^{\oplus}}}{-}CH_3$$

choline

$$CH_3{-}\overset{\overset{\large O}{\|}}{C}{-}O{-}CH_2{-}CH_2{-}\overset{\overset{\large CH_3}{|}}{\underset{\underset{\large CH_3}{|}}{N^{\oplus}}}{-}CH_3$$

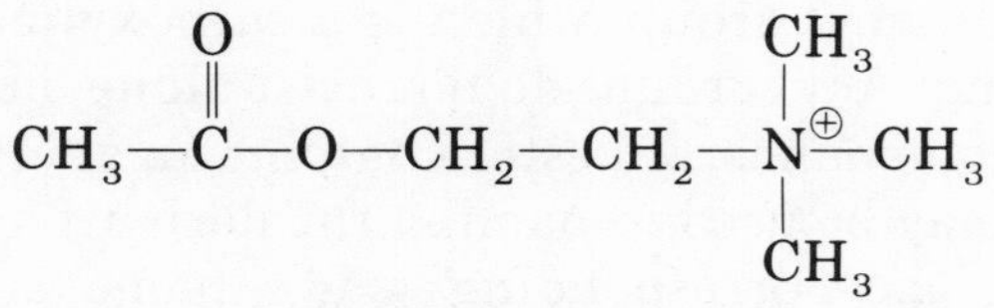

acetyl choline

Acyl groups can be used in naming compounds in much the same way as alkyl radicals are used. While acetyl choline represents another way of naming an ester (the ester of acetic acid and the special alcohol choline), the use of acyl-names is not restricted to esters.

5.1.2 Reactions of Esters

Esters may be hydrolyzed with an acid catalyst. *Hydrolysis* (*hydro* = "water," *lysis* = "breaking") is a reaction in which a covalent bond in a compound, such as an ester, is broken and one molecule of water is added. An OH from the water molecule is joined to one of the broken ends and an H is joined to the other:

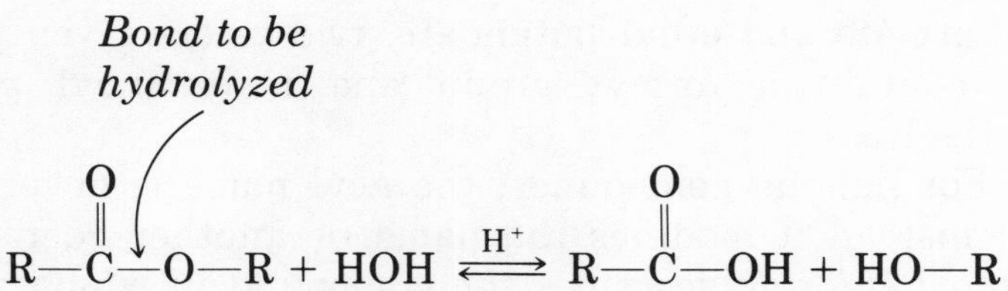

Bond to be hydrolyzed

$$R{-}\overset{\overset{\large O}{\|}}{C}{-}O{-}R + HOH \overset{H^+}{\rightleftharpoons} R{-}\overset{\overset{\large O}{\|}}{C}{-}OH + HO{-}R$$

hydrolysis of an ester

Note that hydrolysis is the reverse of the reaction we called esterification. In fact, the reaction is quite reversible, but the equilibrium lies to the right in aqueous solutions because of the abundance of water, one of the reactants. That is, esters are easily hydrolyzed, but for synthesis water must be removed from the starting compounds in order to displace the reaction to the left. Hydrolysis is a common reaction in biochemistry and many compounds besides esters may be hydrolyzed. Hydrolytic reactions for other functional groups will be discussed later.

Saponification is a name for the general reaction in which esters are cleaved by inorganic bases:

$$\underset{\text{ethyl acetate}}{CH_3\text{—}\overset{\overset{\displaystyle O}{\|}}{C}\text{—}O\text{—}CH_2\text{—}CH_3} + NaOH \longrightarrow \underset{\text{sodium acetate}}{CH_3\text{—}\overset{\overset{\displaystyle O}{\|}}{C}\text{—}O^{\ominus}Na^{\oplus}} + \underset{\text{ethyl alcohol}}{HO\text{—}CH_2\text{—}CH_3}$$

As mentioned before, the salts of long-chain acids are called soaps, which are generally derived from fats. Fats are esters of fatty acids and the special polyhydroxylic alcohol glycerol. Saponification of fats yields soap and glycerol:

$$\underset{\text{a fat}}{\begin{array}{l} CH_3(CH_2)_{16}\overset{\overset{\displaystyle O}{\|}}{C}\text{—}O\text{—}CH_2 \\ \qquad\qquad\qquad\qquad\quad | \\ CH_3(CH_2)_{16}\overset{\overset{\displaystyle O}{\|}}{C}\text{—}O\text{—}CH \\ \qquad\qquad\qquad\qquad\quad | \\ CH_3(CH_2)_{16}\overset{\overset{\displaystyle O}{\|}}{C}\text{—}O\text{—}CH_2 \end{array}} + 3\,NaOH \longrightarrow \underset{\substack{\text{sodium stearate} \\ (\textit{a soap})}}{3\,CH_3(CH_2)_{16}\overset{\overset{\displaystyle O}{\|}}{C}\text{—}O^{\ominus}Na^{\oplus}} + \underset{\text{glycerol}}{\begin{array}{l} HO\text{—}CH_2 \\ \qquad\;\; | \\ HO\text{—}CH \\ \qquad\;\; | \\ HO\text{—}CH_2 \end{array}}$$

Note that a fat molecule has three ester groups and hence has *three acyl groups*. Such fat molecules are called *triacyl*glycerols, or sometimes, less logically, triglycerides. In most native triglycerides the acyl groups are different instead of all being the same as in the example given here. Nevertheless, all three ester groups would be cleaved by NaOH (or KOH) in the same way that ethyl acetate is cleaved.

5.1.3 Flavors and Odors

Fats do not have much odor, but the shorter-chain esters have fruity odors and tastes. Esters are the active agents in many fruit flavors and wines. They are also used in perfumes. One of the components of banana oil is the acetate ester of isoamyl alcohol, or isoamyl acetate. Isoamyl alcohol, a trivial name, is 3-methyl-1-butanol in the Geneva system:

$$CH_3-\overset{\overset{\large O}{\|}}{C}-O-CH_2-CH_2-\overset{\overset{\large CH_3}{|}}{CH}-CH_3$$

isoamyl acetate (trivial)
3-methyl-1-butyl acetate (Geneva)

5.1.4 Physical Properties

Compared to their corresponding acids, esters have much lower water solubilities. This is due to two factors. (1) An alkyl radical (hydrophobic carbon chain) has been added in place of a hydrogen atom. (2) Because the ionizable hydrogen atom of the carboxylic acid is the one replaced by the alkyl radical, esters cannot ionize; consequently the ester function is much less polar than the carboxyl group. Ethyl acetate, whose structure was given earlier, is only partially soluble in water. On the other hand, esters are good solvents for hydrophobic compounds. (Ethyl acetate is the active ingredient in many fingernail-polish removers.)

5.1.5 Lactones

A lactone is a special type of ester that may be formed when a carboxyl group and an alcohol group are present in the same molecule. Such groups may form an *intra*molecular ester linkage to create a lactone. In short, lactones are cyclic esters. As mentioned in the discussion of glucose in Section 3.5, organic compounds have a tendency to form five- and six-membered rings. Consequently carboxylic acids that also have an alcohol group (—OH) at the #4 or #5 position form lactones readily, simply by being heated:

$$CH_3{-}\underset{\displaystyle OH}{\underset{|}{CH}}{-}CH_2{-}CH_2{-}CH_2{-}\overset{\displaystyle O}{\overset{\|}{C}}{-}OH$$

5-hydroxyhexanoic acid

$$\Big\downarrow \text{heat}$$

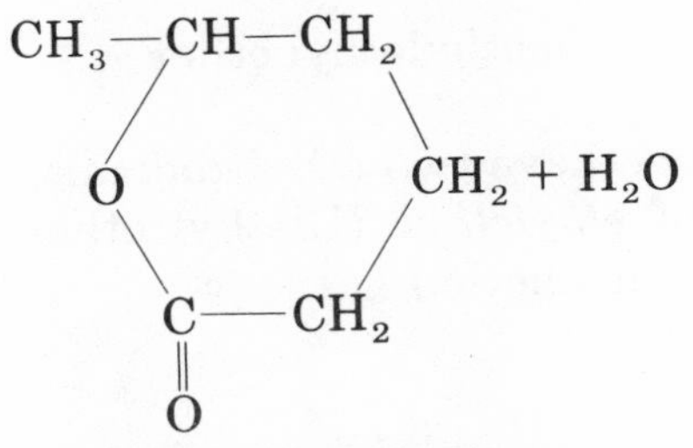

$$\begin{array}{ccccc} & CH_3{-}CH{-}CH_2 & & & \\ \diagup & & \diagdown & & \\ O & & & CH_2 + H_2O & \\ \diagdown & & \diagup & & \\ & \underset{\displaystyle O}{\underset{\|}{C}}{-}CH_2 & & & \end{array}$$

the lactone of 5-hydroxyhexanoic acid

5.2 ETHERS

In effect, an ether represents a combination of two alcohols; thus the general formula is

$$R{-}O{-}R \qquad \text{or} \qquad ROR$$

Again the *R*'s (alkyl radicals) may be different or the same. Diethyl ether may be derived from two molecules of ethyl alcohol:

$$\underset{\text{ethyl alcohol}}{CH_3{-}CH_2{-}OH} + \underset{\text{ethyl alcohol}}{OH{-}CH_2{-}CH_3} \xrightarrow{H_2SO_4}$$

$$\underset{\text{diethyl ether}}{CH_3CH_2{-}O{-}CH_2{-}CH_3} + H_2O$$

The sulfuric acid "pulls" a water molecule from the two alcohol molecules to force the alcohol moieties to combine. The term *moiety* is used frequently to describe simple components that make up a larger molecule. In the example just given, ethyl alcohol is not present in diethyl ether, technically. However, diethyl ether can be derived from two molecules of ethyl alcohol by establishing an ether group and eliminating a molecule of water. Thus the term *alcohol moieties* describes the portions of ethyl ether that were derived from ethyl alcohol, although they certainly are not alcohols after the combination.

Ethers are very insoluble in water and are much like alkanes in many of their properties. Ethers are quite volatile and unreactive. It is difficult to hydrolyze ethers.

To name ethers, the two alkyl radicals are named and a second word, *ether*, is added:

$$CH_3—O—CH_2—CH_3$$

methylethyl ether

In diethyl ether we have two ethyl radicals, and so we write *diethyl* instead of *ethylethyl*. (Diethyl ether is the so-called ether used as an anesthetic.)

EXERCISES

1. Name the following compounds.

(a) $CH_3—CH_2—\overset{\overset{\displaystyle O}{\|}}{C}—O—CH_3$

(b) $CH_3\!\leftarrow\!CH_2\!\rightarrow_{14}\overset{\overset{\displaystyle O}{\|}}{C}—O—CH_2—CH_3$

(c) $CH_3—CH_2—CH_2—\overset{\overset{\displaystyle O}{\|}}{C}—O—CH_2—CH_2—CH_2—CH_3$

2. Write a reaction for the synthesis of the ester methyl acetate.

3. Show a reaction for the saponification of ethyl butanoate.

4. Draw the structures for the following compounds.
(a) ethyl heptanoate
(b) *n*-octyl acetate
(c) methyl butyrate

5. Name the following compounds.
(a) $CH_3—CH_2—O—CH_2—CH_2—CH_3$
(b) $CH_3—O—CH_2—CH_2—CH_2—CH_3$
(c) $CH_3—O—CH_3$

Amines and Amides

6

6.1 AMINES

Amines contain the functional group $—NH_2$, —NHR, or —NRR, and can be regarded as derivatives of ammonia (NH_3). The Lewis structure of the nitrogen atom indicates that it should form three covalent bonds. In ammonia the three bonds are between the nitrogen atom and the three hydrogen atoms. Organic amines are similar except that one, two, or three of the N–H bonds are replaced by N–R bonds. Methylamine is the simplest *primary (1°) amine*:

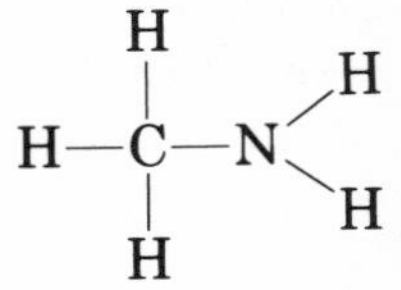

$$H_3C{-}NH_2 \qquad CH_3{-}NH_2 \qquad \text{or} \qquad CH_3NH_2$$

methylamine

Primary amines have *one* of the hydrogen atoms of ammonia replaced by an R-group; therefore their general formula is $R—NH_2$.

Secondary (2°) amines have *two* hydrogen atoms of ammonia replaced by R-groups, and thus have the general formula R_2NH. The two R-groups may be the same or different alkyl radicals. The simplest secondary amine is dimethylamine:

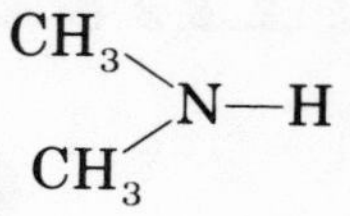

$$\begin{matrix} CH_3 \diagdown \\ \quad N-H \\ CH_3 \diagup \end{matrix} \qquad \text{or} \qquad (CH_3)_2NH$$

dimethylamine

Tertiary (3°) amines have all *three* of the hydrogen atoms of ammonia replaced by R-groups, and so their general formula is R_3N. Again the R-groups may be any alkyl radical whether different or the same. Trimethylamine is the simplest tertiary amine:

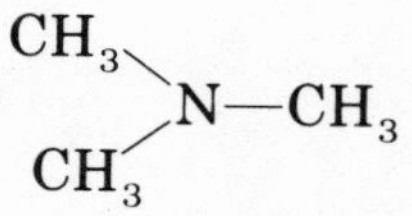

$$\begin{matrix} CH_3 \diagdown \\ \quad N-CH_3 \\ CH_3 \diagup \end{matrix} \qquad \text{or} \qquad (CH_3)_3N$$

trimethylamine

Some examples of mixed tertiary amines are

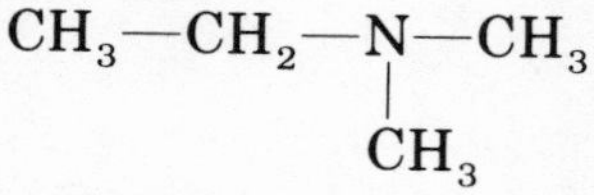

$$\underset{\text{ethyldimethylamine}}{CH_3-CH_2-\underset{\substack{|\\ CH_3}}{N}-CH_3} \qquad \underset{\text{ethylmethylpropylamine}}{CH_3-CH_2-CH_2-\underset{\substack{|\\ CH_3}}{N}-CH_2-CH_3}$$

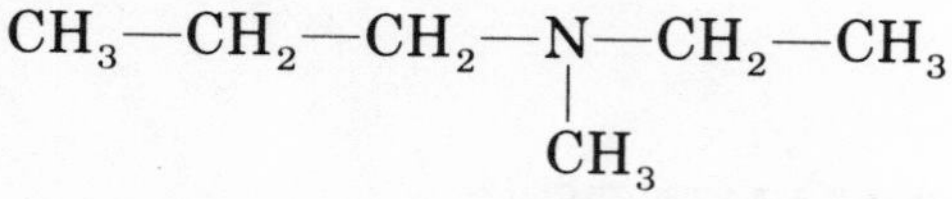

As just demonstrated, amines are named by naming the alkyl radicals attached to the nitrogen atom and then joining these names to the word *amine*. Note that *primary*, *secondary*, and *tertiary* have slightly different meanings for amines than for alcohols. Compare the designations for alcohols (Section 3.1.1) to those of amines.

Amines generally smell fishy, and decaying protein releases amines. Much of the odor of dead animals is due to volatile amines.

A very important property of amines is that they are basic, like ammonia. You will recall that ammonia is basic because hydrogen ions, in water, readily associate with the pair of unshared electrons of the nitrogen atom:

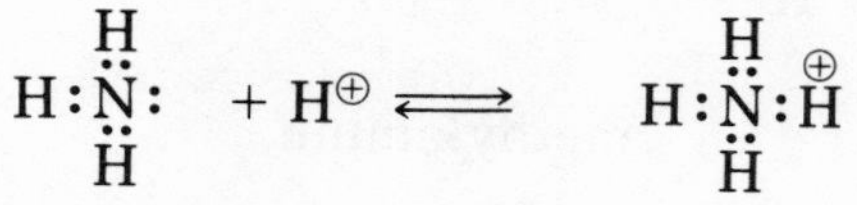

$$\underset{\text{ammonia}}{\begin{matrix} H \\ H:\ddot{N}: \\ H \end{matrix}} + H^{\oplus} \rightleftharpoons \underset{\text{ammonium ion}}{\begin{matrix} H \\ H:\ddot{N}:H^{\oplus} \\ \ddot{H} \end{matrix}}$$

In ammonia the nitrogen atom supplies five electrons and each of the three hydrogen atoms supplies one. The proton simply latches on to the free unshared pair of electrons. In the ammonium ion the "freeloader proton" cannot be distinguished from the other hydrogen atoms that donated an electron.

The stability of the ammonium ion can be partially explained by an idea closely related to resonance. Consider the following reversible reactions:

$$\begin{array}{ccccc} & & H:\overset{\cdot\cdot}{\underset{\cdot\cdot}{N}}:H + H^{\oplus} & & \\ & & H & & \\ & & \Updownarrow & & \\ & H & H & H & \\ H^{\oplus} + :\overset{\cdot\cdot}{N}:H & \rightleftharpoons & H:\overset{\cdot\cdot}{\underset{\cdot\cdot}{N}}{}^{\oplus}:H & \rightleftharpoons & H:\overset{\cdot\cdot}{\underset{\cdot\cdot}{N}}: + H^{\oplus} \\ H & & H & & H \\ & & \Updownarrow & & \\ & & H & & \\ & & H:\underset{\cdot\cdot}{\overset{\cdot\cdot}{N}}:H + H^{\oplus} & & \end{array}$$

The more stable an ammonium ion is, the more basic it will be, because it holds the proton ($H^{\oplus}$) more strongly if it is more stable. As you will recall, ammonia does not need a strong acid to donate a proton to it. It can take the proton from water and thus generate an excess of $OH^{\ominus}$, which explains why ammonia is basic:

$$NH_3 + H_2O \rightleftharpoons NH_4^{\oplus} + OH^{\ominus}$$

About 0.4% of the ammonia dissolved in water will exist in the $NH_4^{\oplus}$ form. Primary, secondary, and tertiary amines will react similarly. This is illustrated with methylamine:

$$\begin{array}{c} H\ H \\ H:\overset{\cdot\cdot}{\underset{\cdot\cdot}{C}}:\overset{\cdot\cdot}{\underset{\cdot\cdot}{N}}: \\ H\ H \end{array} + H:\overset{\cdot\cdot}{\underset{\cdot\cdot}{O}}:H \longrightarrow \begin{array}{c} H\ H \\ H:\overset{\cdot\cdot}{\underset{\cdot\cdot}{C}}:\overset{\cdot\cdot}{\underset{\cdot\cdot}{N}}{}^{\oplus}:H \\ H\ H \end{array} + :\overset{\cdot\cdot}{\underset{\cdot\cdot}{O}}:H^{\ominus}$$

If amines are treated with an acid, an ammonium salt is obtained. This general reaction is illustrated with methylamine:

$$\underset{\text{methylamine}}{CH_3{-}NH_2} + HCl \longrightarrow \underset{\text{methylammonium chloride}}{[CH_3{-}NH_3]^{\oplus}Cl^{\ominus}}$$

Methylammonium chloride may be called "the hydrochloride salt of methyl amine," because hydrochloric acid was added to methylamine to form the salt. The $CH_3{-}NH_3^{\oplus}$ is called the methylammonium ion.

For naming, note that by the addition of a proton, *methylamine* becomes a *methylammonium ion.* Thus the alkyl radicals attached to the nitrogen atom are named the same in both types of compounds, but if the molecule is neutral it is called an amine and if it carries a positive charge it is called an ammonium ion. Salts of ammonium ions are named by: (1) naming the alkyl radicals, (2) joining these names to the word *ammonium* (to indicate that it exists as a cation), and (3) following this with the name of the anion associated with the cation. Do you see each of these features in the following compound?

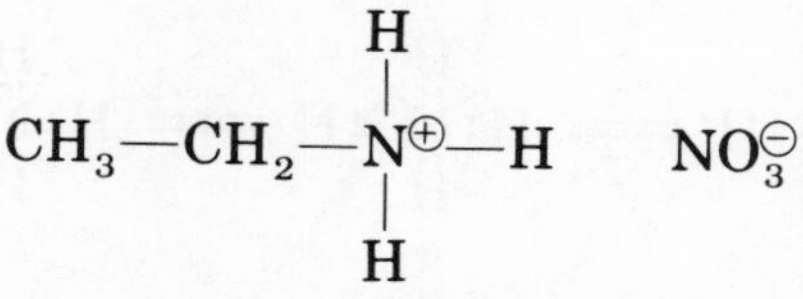

$$CH_3{-}CH_2{-}\overset{\displaystyle H}{\underset{\displaystyle H}{\overset{|}{\underset{|}{N^{\oplus}}}}}{-}H \quad NO_3^{\ominus}$$

ethylammonium nitrate

Primary, secondary, and tertiary amines may form ammonium ions:

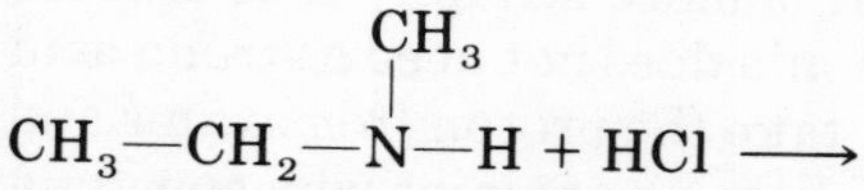

$$CH_3{-}CH_2{-}\overset{\displaystyle CH_3}{\overset{|}{N}}{-}H + HCl \longrightarrow$$

methylethylamine

$$CH_3{-}CH_2{-}\overset{\displaystyle CH_3}{\underset{\displaystyle H}{\overset{|}{\underset{|}{N^{\oplus}}}}}{-}H \quad Cl^{\ominus}$$

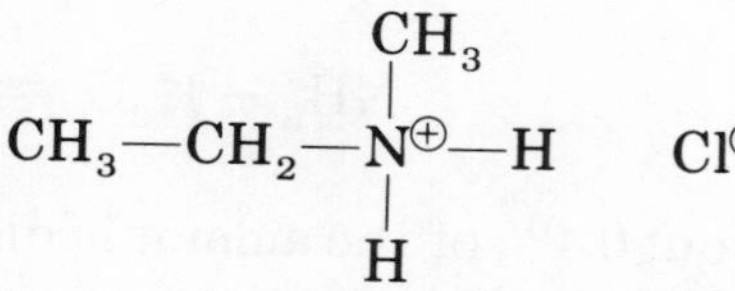

methylethylammonium chloride

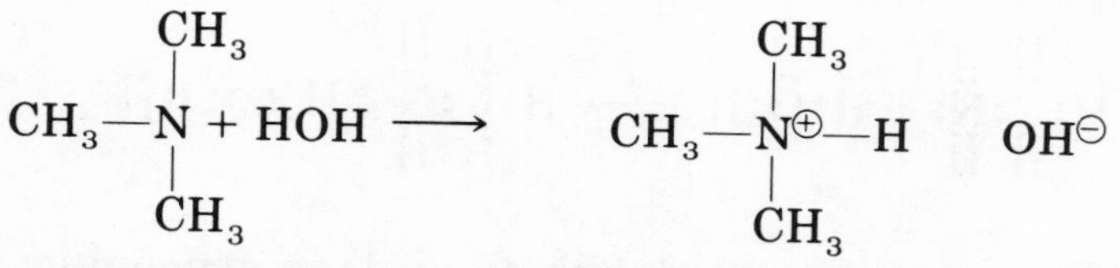

$$CH_3{-}\overset{\displaystyle CH_3}{\underset{\displaystyle CH_3}{\overset{|}{\underset{|}{N}}}} + HOH \longrightarrow CH_3{-}\overset{\displaystyle CH_3}{\underset{\displaystyle CH_3}{\overset{|}{\underset{|}{N^{\oplus}}}}}{-}H \quad OH^{\ominus}$$

trimethylamine — trimethylammonium hydroxide

Surprise—*quaternary (4°)* ammonium ions exist. They have *four* alkyl radicals substituted for the four hydrogen atoms in the ammonium ion. Like hydrogen, one R-group does not donate an electron; it uses the unshared pair of electrons from the nitrogen atom. This causes a *permanent* positive charge to exist on the resulting molecule:

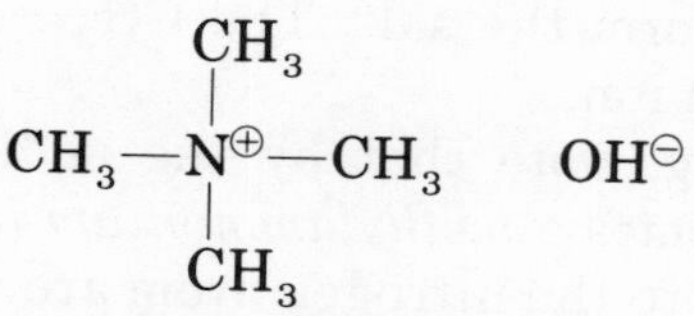

$$CH_3{-}\overset{\displaystyle CH_3}{\underset{\displaystyle CH_3}{\overset{|}{\underset{|}{N^{\oplus}}}}}{-}CH_3 \quad OH^{\ominus}$$

tetramethylammonium hydroxide

An example of a compound involved in nerve-impulse conduction is acetyl choline, which was discussed in Section 5.1.1 in connection with esters and acyl groups:

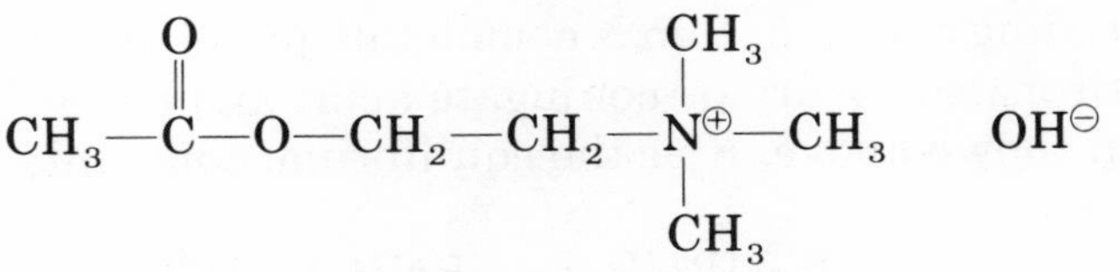

$$CH_3-\overset{\overset{\displaystyle O}{\|}}{C}-O-CH_2-CH_2-\underset{\underset{\displaystyle CH_3}{|}}{\overset{\overset{\displaystyle CH_3}{|}}{N^{\oplus}}}-CH_3 \qquad OH^{\ominus}$$

acetyl choline hydroxide

In addition to the ester function, acetyl choline has a *quaternary ammonium group*.

While in our discussion quaternary ammonium compounds have been illustrated with methyl groups, various R-groups, whether the same or different, may be involved.

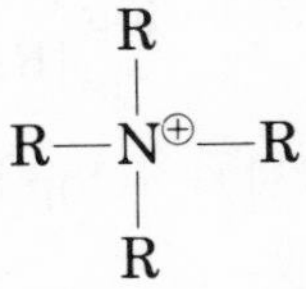

$$R-\underset{\underset{\displaystyle R}{|}}{\overset{\overset{\displaystyle R}{|}}{N^{\oplus}}}-R$$

a quaternary ammonium ion

6.1.1 Ionization of Amines

Amines should be thought of as organic bases; so it is important to consider their ionization reactions in detail. Consider the following:

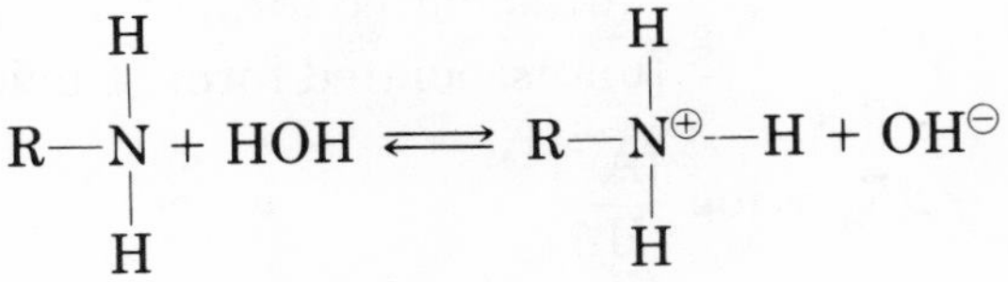

$$R-\underset{\underset{\displaystyle H}{|}}{\overset{\overset{\displaystyle H}{|}}{N}} + HOH \rightleftharpoons R-\underset{\underset{\displaystyle H}{|}}{\overset{\overset{\displaystyle H}{|}}{N^{\oplus}}}-H + OH^{\ominus}$$

This reaction is an ionization reaction that generates $OH^{\ominus}$. Thus the equilibrium constant for such a reaction would be K_b because the subscript b indicates the ionization of a base whereas the subscript a stands for the ionization of an acid. The $RNH_3^{\oplus}$ can be treated as an acid according to the following reaction:

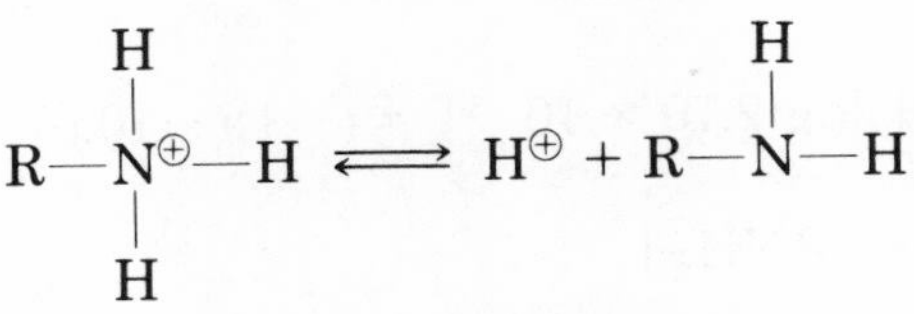

$$R-\underset{\underset{\displaystyle H}{|}}{\overset{\overset{\displaystyle H}{|}}{N^{\oplus}}}-H \rightleftharpoons H^{\oplus} + R-\overset{\overset{\displaystyle H}{|}}{N}-H$$

This is the reverse of the previous reaction if the ionization of water were taken into account. For our purposes the latter equation will be used, but it should be realized that the $R—NH_3^{\oplus}$ will not dissociate appreciably except on the basic side of the pH scale (above pH 7). The ammonium ion is the conjugate acid of the base, which is the neutral amine. Thus the preceding manipulation is completely justified on the basis of the Brønsted theory of conjugate acids and bases. For the reaction shown above, K_a is the equilibrium constant:

$$K_a = \frac{[H^{\oplus}][RNH_2]}{[RNH_3^{\oplus}]} = \frac{[RNH_2]}{[RNH_3^{\oplus}]} \cdot \frac{[H^{\oplus}]}{1}$$

Take the log of both sides of the equation:

$$\log K_a = \log \frac{[RNH_2]}{[RNH_3^{\oplus}]} + \log [H^{\oplus}]$$

$$-\log [H^+] + \log K_a = \log \frac{[RNH_2]}{[RNH_3^{\oplus}]}$$

$$-\log [H^+] = -\log K_a + \log \frac{[RNH_2]}{[RNH_3^{\oplus}]}$$

Recall the definitions of pH and pK_a:

$$pH = pK_a + \log \frac{[RNH_2]}{[RNH_3^{\oplus}]}$$

The equation is similar to that developed for acetic acid and acids in general. It is the Henderson-Hasselbalch equation, which may be written in more general terms:

$$pH = pK_a + \log \frac{[\text{dissociated form of acid}]}{[\text{undissociated form of acid}]}$$

$$= pK_a + \log \frac{[A^{\ominus}]}{[HA]}$$

or

$$pH = pK_a + \log \frac{[\text{conjugate base}]}{[\text{conjugate acid}]}$$

Conjugate acids and bases do not have to be neutral or charged; they may be either.

The K_a for methylamine is 2.70×10^{-11}, and

$$pK_a = (-1)\log 2.70 \times 10^{-11} = (-1)(-10.57) = 10.57$$

When $[RNH_2] = [RNH_3^{\oplus}]$

$$pH = 10.57 + \log \frac{[RNH_2]}{[RNH_3^{\oplus}]} = 10.57 + 0$$

Thus at a pH of 10.57 the concentrations of the dissociated and undissociated forms of the methylammonium ion are equal.

What would the ratio of the two forms be if the pH were 7? The answer is as follows:

$$7 = 10.57 + \log \frac{[RNH_2]}{[RNH_3^{\oplus}]}$$

$$-3.57 = \log \frac{[RNH_2]}{[RNH_3^{\oplus}]}$$

$$0.000269 = \frac{[RNH_2]}{[RNH_3^{\oplus}]}$$

This specifies that the majority of the amine is in the acid ($R—NH_3^{\oplus}$) form. The reciprocal of the ratio may be calculated:

$$3.57 = \log \frac{[RNH_3^{\oplus}]}{[RNH_2]}$$

$$\frac{[RNH_3^{\oplus}]}{[RNH_2]} = 3715$$

which means that there are 3715 molecules of $RNH_3^{\oplus}$ for each RNH_2 molecule at a pH of 7.

6.1.2 Solubility of Amines

The substitution of a polar functional group, such as a hydroxyl group, for a hydrogen atom increases the solubility of an organic molecule in water. However, the substitution of an ionized group, such as a carboxylate group ($—COO^{\ominus}$) or an ammonium group ($—NH_3^{\oplus}$), has a more pronounced effect in enhancing solubility. You should not be surprised to learn that an organic ammonium ion is much more soluble in water than the corresponding neutral amine. The amine form is more soluble in hydrophobic organic solvents, such as hexane or diethyl ether.

6.1.3 Hydrogen Bonding

The hydrogen atoms attached to amines form hydrogen bonds readily. In fact, hydrogen atoms attached to nitrogen atoms in other functional groups are subject to hydrogen bonding. Such groups are instrumental in determining and maintaining the topology of large molecules such as nucleic acids (RNA and DNA) and proteins. Hydrogen bonding between a ketone and

an amine is given as

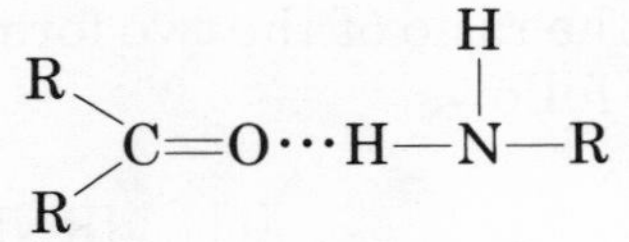

$$\begin{array}{c} \mathrm{R} \\ \diagdown \\ \\ \diagup \\ \mathrm{R} \end{array}\!\!\mathrm{C{=}O\cdots H{-}}\begin{array}{c}\mathrm{H}\\ | \\ \mathrm{N} \end{array}\!\mathrm{{-}R}$$

The relationships between the participating hydrogen atom, the oxygen atom, and nitrogen atoms are similar to those involved in hydrogen bonding of water (Section 3.1.5) and of ketones (Section 3.6).

6.2 AMIDES

Amides are derivatives of carboxylic acids in which the —OH-group of the carboxyl group has been replaced by

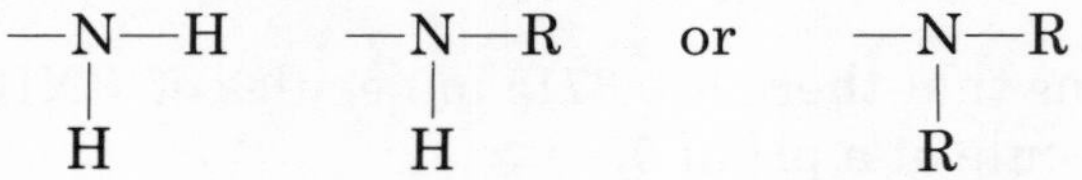

$$\mathrm{{-}\underset{\underset{H}{|}}{N}{-}H} \qquad \mathrm{{-}\underset{\underset{H}{|}}{N}{-}R} \qquad \text{or} \qquad \mathrm{{-}\underset{\underset{R}{|}}{N}{-}R}$$

For example, acetic acid can be converted to acetamide:

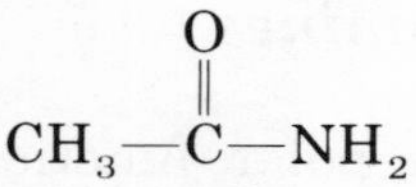

$$\mathrm{CH_3{-}\overset{\overset{O}{\|}}{C}{-}NH_2}$$

acetamide

Like amines, amides may be classified. Acetamide is a primary amide. The general structures for primary (1°), secondary (2°), and tertiary (3°) amides are given by the following:

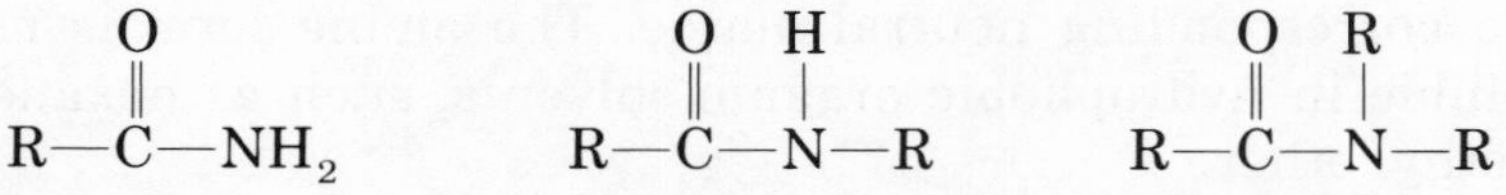

$$\mathrm{R{-}\overset{\overset{O}{\|}}{C}{-}NH_2} \qquad \mathrm{R{-}\overset{\overset{O}{\|}}{C}{-}\overset{\overset{H}{|}}{N}{-}R} \qquad \mathrm{R{-}\overset{\overset{O}{\|}}{C}{-}\overset{\overset{R}{|}}{N}{-}R}$$

a primary amide *a secondary amide* *a tertiary amide*

The distinction between these amides parallels that for amines. However, one of the hydrogen-atom replacements on ammonia (NH_3) is an acyl group ($\mathrm{R{-}\overset{\overset{O}{\|}}{C}{-}}$) rather than an R-group. A secondary amide has two of the hydrogen atoms of ammonia replaced, one by an acyl group and one by an R-group, while a tertiary amide has one acyl group and two R-groups.

Substitution of an acyl group ($R-\overset{O}{\overset{\|}{C}}-$) for an alkyl radical (R—) or a hydrogen atom on a nitrogen atom has a profound effect on the chemical nature of the nitrogen atom. Unlike with amines and ammonia, the nitrogen atoms of amides *do not* behave as bases, and *quaternary amides are not encountered*.

Names of amides are derived from their respective acids. For example, in the following compound butanoic acid is the corresponding acid. The *-oic* and *acid* are dropped and *-amide* is added as a suffix to the first word, to yield *butanamide*:

$$CH_3-CH_2-CH_2-\overset{O}{\overset{\|}{C}}-NH_2$$

butanamide

For a secondary amide the name of the alkyl radical attached to the nitrogen atom is indicated at the beginning of the name with a capital *N*:

$$CH_3-CH_2-CH_2-CH_2-\overset{O}{\overset{\|}{C}}-\overset{H}{\overset{|}{N}}-CH_3$$

N-methylpentanamide

The *N*- in front of the *methyl-* specifies that the methyl group is substituted on the nitrogen atom. This is necessary because a methyl group could be attached to one of the carbon atoms, but if that were the case, a number would have preceded the *methyl-* to indicate which carbon atom bore the group. Tertiary amides are named similarly:

$$CH_3-CH_2-CH_2-CH_2-\overset{O}{\overset{\|}{C}}-\overset{CH_3}{\overset{|}{N}}-CH_2-CH_3$$

N-methyl, *N*-ethylpentanamide

$$CH_3(CH_2)_3\overset{O}{\overset{\|}{C}}-\overset{CH_3}{\overset{|}{N}}-CH_3$$

N,N-dimethylpentanamide

Note that one capital *N* is used for each alkyl radical substituted on the nitrogen atom of the amide. Care should be exercised in the use of the letter *N* for naming organic compounds. A capital *N* means something much different than a lowercase *n*, which stands for *normal* and is used to indicate that an alkyl radical has an unbranched chain.

The term *N-substitution* is used to indicate that an alkyl group has been substituted for a hydrogen atom in *amines* as well as in *amides*. Consequently the three compounds just presented may be called *N*-substituted amides.

Amides may be hydrolyzed with strong acid as a catalyst:

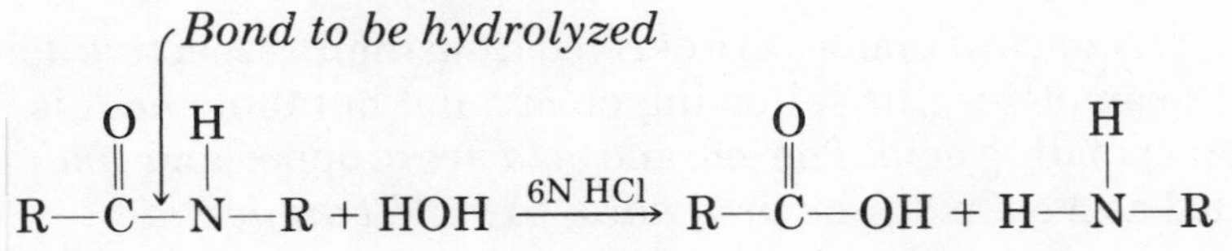

Bond to be hydrolyzed

$$R{-}\overset{\overset{\displaystyle O}{\|}}{C}{\downarrow}\overset{\overset{\displaystyle H}{|}}{N}{-}R + HOH \xrightarrow{6N\ HCl} R{-}\overset{\overset{\displaystyle O}{\|}}{C}{-}OH + H{-}\overset{\overset{\displaystyle H}{|}}{N}{-}R$$

A secondary amide is shown in the reaction, but primary and tertiary amides are hydrolyzed similarly. All three classes of amides yield a carboxylic acid; ammonia, a primary amine, and a secondary amine are the other hydrolytic products for primary, secondary, and tertiary amides, respectively. The bond that is hydrolyzed (between the acyl group and the nitrogen atom) is called an *amide bond*. For secondary amides, such a bond is also called a *peptide bond* because secondary amide bonds are present in peptides and proteins, which are made by combining amino acids.

Amides may form hydrogen bonds in much the same way that amines do. Hydrogen bonding between a hydrogen atom on a secondary amide group and a carbonyl group (oxygen double bonded to carbon) is instrumental in maintaining certain structural forms of protein molecules:

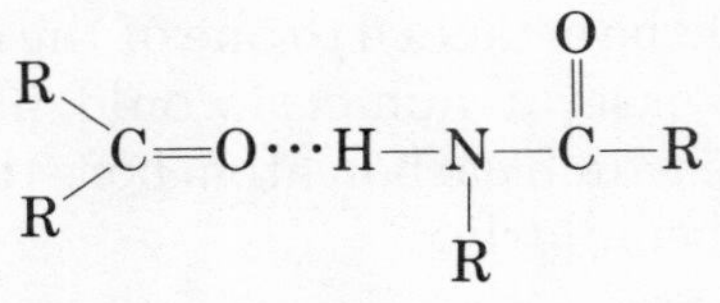

$$\begin{matrix} R \\ R \end{matrix}\!\!>\!C{=}O\cdots H{-}\underset{\underset{\displaystyle R}{|}}{N}{-}\overset{\overset{\displaystyle O}{\|}}{C}{-}R$$

hydrogen bonding by a secondary amide and a carbonyl group

EXERCISES

1. Indicate whether each of the following is a primary, secondary, tertiary, or quaternary amine.
   (a) $CH_3{-}CH_2{-}CH_2{-}NH_2$
   (b) 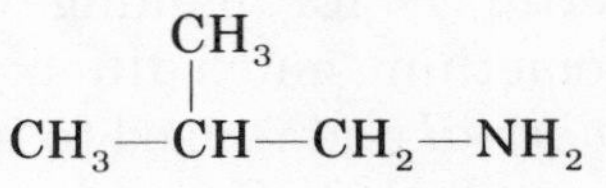 $CH_3{-}\underset{}{\overset{\overset{\displaystyle CH_3}{|}}{C}H}{-}CH_2{-}NH_2$

$$\begin{array}{c} CH_3 \\ | \\ CH_3\text{—}CH_2\text{—}N\text{—}H \end{array}$$
(c)

(f)

$$\begin{array}{c} CH_3 \\ | \\ CH_3\text{—}CH_2\text{—}N\text{—}CH_3 \end{array}$$
(d)

(e) $CH_3\text{—}CH_2\text{—}NH\text{—}CH_3$

$$\begin{array}{c} CH_3 \\ | \\ CH_3\text{—}CH_2\text{—}N^{\oplus}\text{—}CH_2\text{—}CH_3 \\ | \\ CH_3 \end{array}$$
(f)

2. Draw the Lewis structure of the ammonium ion ($NH_4^{\oplus}$), showing which electrons were donated by the hydrogen atoms and which by the nitrogen atom.

3. Name the following compounds.
   (a) $CH_3\text{—}CH_2\text{—}CH_2\text{—}NH_2$
   (b) $CH_3\text{—}CH_2\text{—}NH\text{—}CH_2\text{—}CH_3$
   (c) $\begin{array}{c} CH_3 \\ | \\ CH_3\text{—}N\text{—}CH_3 \end{array}$
   (d) $CH_3\text{—}CH_2\text{—}CH_2\text{—}CH_2\text{—}NH_2$

4. Show how ethylamine reacts with water to form the ethylammonium ion ($Et\text{—}NH_3^{\oplus}$). Incidentally, sometimes Et is used to indicate an ethyl radical, and Me to indicate a methyl radical.

5. Write the reaction that illustrates the ethylammonium ion's behavior as an acid.

6. If the pK_a of ethylamine (actually $CH_3\text{—}CH_2\text{—}NH_3^{\oplus}$) is 10.67, calculate the pH of a solution that contains 10 mM $CH_3\text{—}CH_2\text{—}NH_3^{\oplus}$ and 1 mM $CH_3\text{—}CH_2\text{—}NH_2$.

7. (a) From the pK_a of ethylamine given in question 6, calculate the ratio of two forms [dissociated]/[undissociated] of ethylamine when the pH of a solution is 9.98.
   (b) Then calculate the concentration of the ethylammonium ion ($Et\text{—}NH_3^{\oplus}$) at that pH when the total concentration of both forms of ethylamine is equal to 0.1 M. That is, $[Et\text{—}NH_3^{\oplus}] + [Et\text{—}NH_2] = 0.1$ M.

8. Name the following compounds.
   (a) $\begin{array}{c} \qquad\qquad\qquad\quad O \\ \qquad\qquad\qquad\quad \| \\ CH_3\text{—}CH_2\text{—}CH_2\text{—}C\text{—}NH_2 \end{array}$
   (b) $\begin{array}{c} \qquad\qquad O \\ \qquad\qquad \| \\ CH_3\text{—}CH_2\text{—}C\text{—}NH\text{—}CH_3 \end{array}$
   (c) $\begin{array}{c} \qquad\qquad O \;\; CH_2\text{—}CH_3 \\ \qquad\qquad \| \quad | \\ CH_3(CH_2)_4C\text{—}N\text{—}CH_2\text{—}CH_2\text{—}CH_3 \end{array}$

9. Show the reaction for the hydrolysis of *N*-ethylpentanamide with a strong acid catalyst (6N HCl).

10. Draw the structures for the following compounds.
(a) *N*,*N*-Dimethylhexanamide
(b) Propionamide
(c) *N*-Ethylacetamide

Anhydrides, and Sulfur- and Phosphate-Containing Compounds

7.1 ANHYDRIDES

Anhydrides are produced by removing a molecule of water from two carboxylic acids. The formation of acetic anhydride is an example.

$$\begin{array}{c} CH_3-C(=O)-OH \\ + \\ CH_3-C(=O)-OH \end{array} \longrightarrow \begin{array}{c} CH_3-C(=O) \\ \quad \diagdown O + H_2O \\ CH_3-C(=O) \end{array}$$

two molecules of acetic acid — acetic anhydride

The general structure of an anhydride is

$$\begin{array}{c} R-C(=O) \\ \quad \diagdown O \\ R-C(=O) \end{array} \quad \text{or} \quad R-\overset{\overset{\displaystyle O}{\|}}{C}-O-\overset{\overset{\displaystyle O}{\|}}{C}-R$$

an acid anhydride

To name an anhydride, the acid's name is used, but the word *acid* is replaced with *anhydride.*

Anhydrides are very reactive and easily hydrolyzed to acids, the reverse of the reaction just shown. In fact, rigorous conditions, including removal of water, are necessary to make anhydrides from acids. Anhydrides are useful for synthesis of organic compounds such as esters or amides:

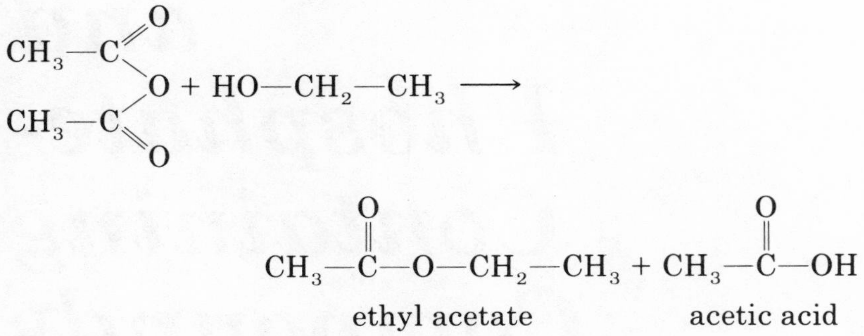

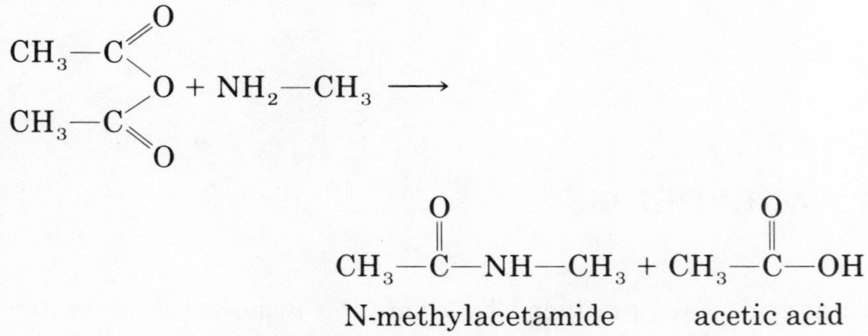

$$(CH_3CO)_2O + HO{-}CH_2{-}CH_3 \longrightarrow \underset{\text{ethyl acetate}}{CH_3{-}\overset{\overset{\displaystyle O}{\|}}{C}{-}O{-}CH_2{-}CH_3} + \underset{\text{acetic acid}}{CH_3{-}\overset{\overset{\displaystyle O}{\|}}{C}{-}OH}$$

$$(CH_3CO)_2O + NH_2{-}CH_3 \longrightarrow \underset{\text{N-methylacetamide}}{CH_3{-}\overset{\overset{\displaystyle O}{\|}}{C}{-}NH{-}CH_3} + \underset{\text{acetic acid}}{CH_3{-}\overset{\overset{\displaystyle O}{\|}}{C}{-}OH}$$

Both of these reactions are general reactions. Various anhydrides and various alcohols may be used to make different esters. Likewise, various anhydrides and amines may be used to make other amides.

Carboxylic acids can form what are called mixed anhydrides, where the two acids are different. Also, cyclic anhydrides can be formed from some dicarboxylic acids such as glutaric acid, $HO_2C{+}CH_2{+}_3CO_2H$:

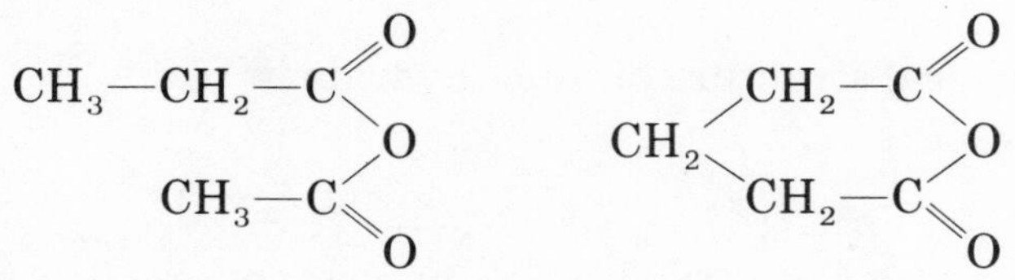

$$CH_3{-}CH_2{-}C(=O){-}O{-}C(=O){-}CH_3 \qquad\qquad CH_2\!\left<\begin{matrix}CH_2{-}C(=O)\\ CH_2{-}C(=O)\end{matrix}\right>\!O$$

propionic acetic anhydride (*a mixed anhydride*)

glutaric anhydride (*a cyclic anhydride*)

Analogous compounds involving an inorganic acid and a carboxylic acid can be formed. As in the case of organic acid anhydrides, a molecule of water is removed and the two acids are joined:

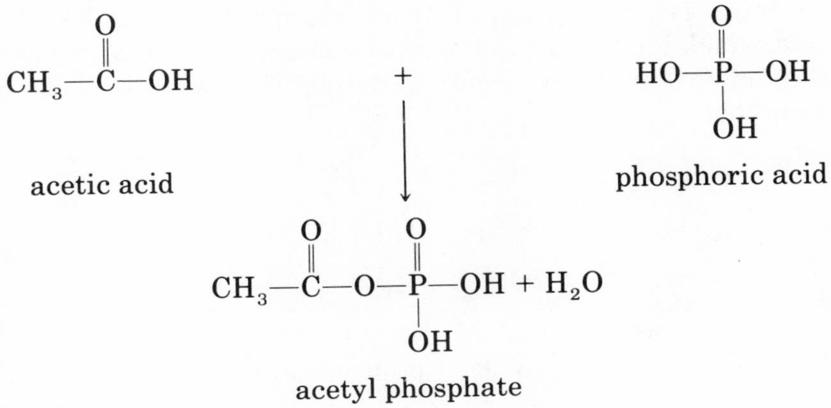

$$CH_3{-}\overset{\overset{O}{\|}}{C}{-}OH \quad + \quad HO{-}\overset{\overset{O}{\|}}{\underset{\underset{OH}{|}}{P}}{-}OH$$

acetic acid — phosphoric acid

$$\downarrow$$

$$CH_3{-}\overset{\overset{O}{\|}}{C}{-}O{-}\overset{\overset{O}{\|}}{\underset{\underset{OH}{|}}{P}}{-}OH + H_2O$$

acetyl phosphate

7.2 SULFUR-CONTAINING COMPOUNDS

Sulfur belongs to the same chemical group as oxygen (see a periodic table), and so it should not be surprising to find that most of the functional groups that contain sulfur are analogous to functional groups that contain oxygen. You may recall that hydrogen sulfide, H_2S, is analogous to water, H_2O.

Mercaptans are *sulfur alcohols* or *thio alcohols*:

$$CH_3{-}CH_2{-}CH_2{-}CH_2{-}SH$$

n-butyl mercaptan

The general formula for a mercaptan is R—SH and the mercaptan group —SH is called a thiol. The *thio-* prefix indicates a substitution of sulfur for an oxygen; hence a thiol is an alcohol (indicated by the *-ol*) where sulfur replaces oxygen. Incidentally, the mercaptan shown here is the active agent in skunk odor. Many sulfur-containing compounds, particularly mercaptans, are vile smelling.

Sulfides are analogous to ethers and may be called thioethers:

$$CH_3{-}CH_2{-}S{-}CH_2{-}CH_3$$

diethyl sulfide

Disulfides are analogous to peroxides, which have not been discussed previously:

$$R{-}S{-}S{-}R' \qquad R{-}O{-}O{-}R'$$

a disulfide — *a peroxide*

Both disulfides and peroxides are reactive (unstable). The disulfide linkage, —S—S—, is important in the structure of proteins. Two thiol groups may be oxidized to form a disulfide, and a disulfide linkage can be reduced to give two thiol groups. An important amino acid, cysteine, is involved in such reactions:

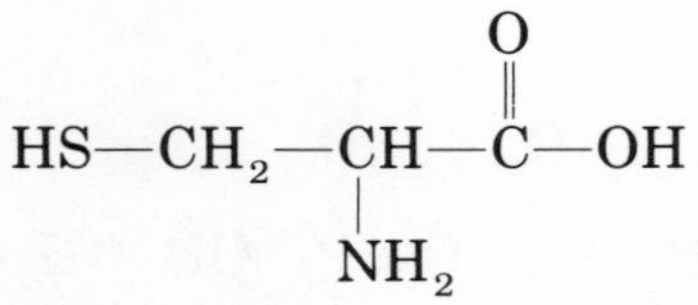

$$HS-CH_2-\underset{\displaystyle NH_2}{\underset{|}{CH}}-\overset{\displaystyle O}{\overset{\|}{C}}-OH$$

cysteine
(2-amino-3-mercaptopropanoic acid)

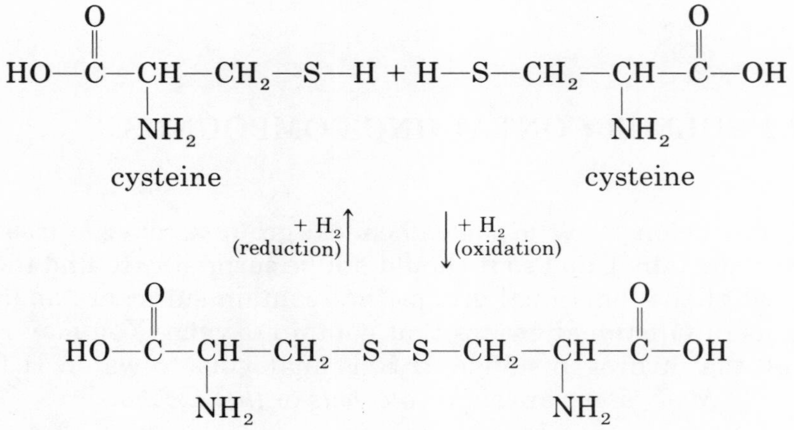

$$HO-\overset{\displaystyle O}{\overset{\|}{C}}-\underset{\displaystyle NH_2}{\underset{|}{CH}}-CH_2-S-H + H-S-CH_2-\underset{\displaystyle NH_2}{\underset{|}{CH}}-\overset{\displaystyle O}{\overset{\|}{C}}-OH$$

cysteine cysteine

$$\underset{\text{(reduction)}}{+H_2}\uparrow \qquad \downarrow\underset{\text{(oxidation)}}{+H_2}$$

$$HO-\overset{\displaystyle O}{\overset{\|}{C}}-\underset{\displaystyle NH_2}{\underset{|}{CH}}-CH_2-S-S-CH_2-\underset{\displaystyle NH_2}{\underset{|}{CH}}-\overset{\displaystyle O}{\overset{\|}{C}}-OH$$

cystine

Note the slight difference in spelling between the disulfide and the mercaptans in this example.

Thiolesters have the general formula

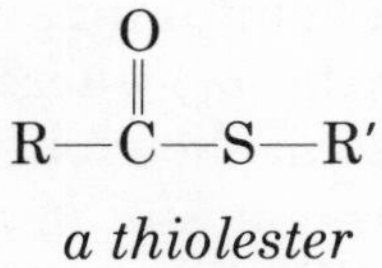

$$R-\overset{\displaystyle O}{\overset{\|}{C}}-S-R'$$

a thiolester

The only structural difference between a thiolester and a regular ester is that a mercaptan moiety has replaced the alcohol moiety.

A complex molecule that contains pantothenic acid (a B-vitamin) and a terminal sulfhydryl group (another name for a thiol or an —SH-group) is called coenzyme A. Although there are many different functional groups in the molecule, the sulfhydryl group is the group that participates most directly in biochemical reactions. Since R—SH represents the general structure of a mercaptan or thiol, the abbreviation

CoA—SH is logical and highlights the importance of the —SH-group. The thiolester linkage "activates" the acyl group, meaning that the acyl portion of the molecule is more reactive. The thiolester formed from coenzyme A and acetic acid ($CoA—S—\overset{\displaystyle O}{\overset{\|}{C}}—CH_3$) is called *acetyl coenzyme A*, *acetyl CoA*, or sometimes *active acetate*.

7.3 PHOSPHATE-CONTAINING COMPOUNDS

From introductory chemistry, you may recall that ortho, or ordinary, phosphoric acid, H_3PO_4, has the following structure:

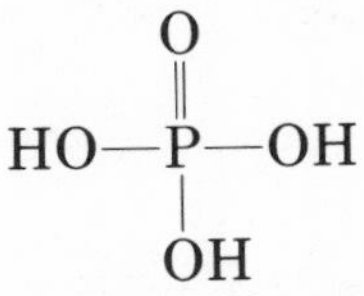

$$HO—\underset{\displaystyle OH}{\underset{|}{\overset{\displaystyle O}{\overset{\|}{P}}}}—OH$$

ortho phosphoric acid

There is considerable similarity between phosphoric acid and carboxylic acids but phosphoric acid has three ionizable groups, and so the ionic forms are $H_2PO_4^{\ominus}$, HPO_4^{-2}, and PO_4^{-3}.

When ionized with a -1, -2, or -3 charge, the acid is called a phosphate ion. If attached to an organic compound, it is called a phosphate or phosphoryl group. Like carboxylic acids, phosphoric acid and a alcohol can form an ester. Such esters are called phosphate esters:

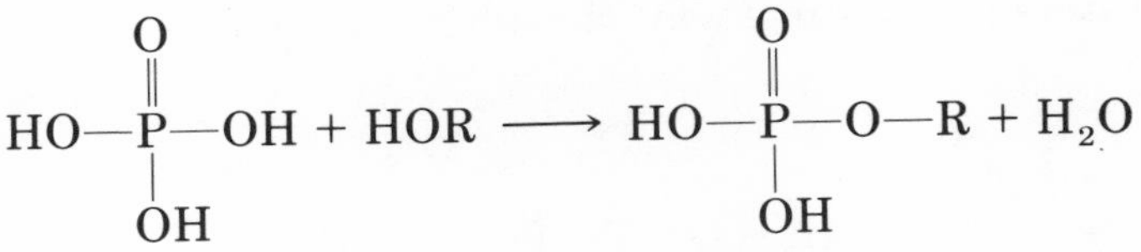

$$HO—\underset{\displaystyle OH}{\underset{|}{\overset{\displaystyle O}{\overset{\|}{P}}}}—OH + HOR \longrightarrow HO—\underset{\displaystyle OH}{\underset{|}{\overset{\displaystyle O}{\overset{\|}{P}}}}—O—R + H_2O$$

a phosphate ester

In naming phosphate esters, the group replacing one of the hydrogen atoms on the phosphate group is named and followed by the word *phosphate*. In the preceding example, if R = $CH_3—CH_2—$, then the phosphate ester would be called *ethyl phosphate*.

Many biological compounds have phosphate groups in their structures. Consider, for example,

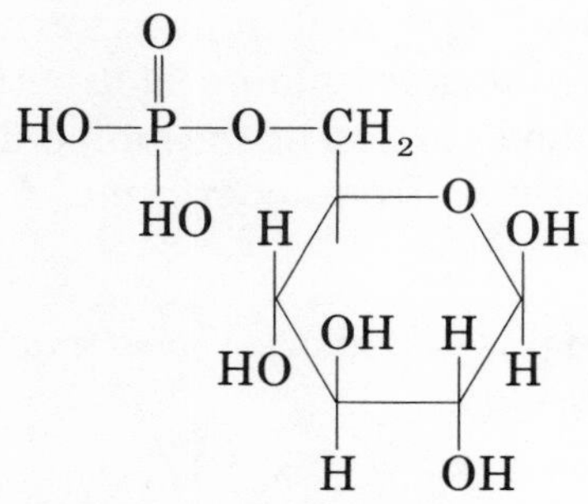

```
         O
         ‖
HO—P—O—CH2
         |
        HO   H  /|——O      OH
               |           \  |
               |\ OH  H  /|
            HO  \|——|/  H
                 |      |
                 H     OH
```

glucose 6-phosphate

We can see that glucose-6-phosphate is a phosphate ester of glucose, shown in Section 3.5. Since there are five hydroxyl groups on glucose, phosphate groups could be attached to any one, or all of them, but in this example the phosphate group is attached to the #6 carbon atom of glucose.

Carboxylic acids are not the only acids that can form anhydrides. The anhydride of two phosphoric acid molecules is called pyrophosphoric acid:

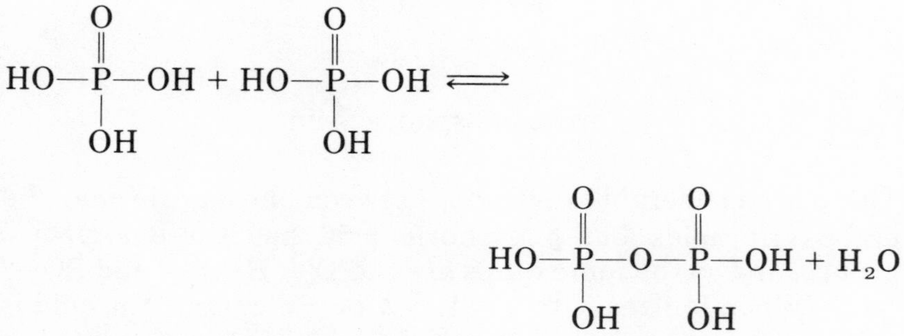

$$HO-\overset{\overset{\displaystyle O}{\|}}{\underset{\underset{\displaystyle OH}{|}}{P}}-OH + HO-\overset{\overset{\displaystyle O}{\|}}{\underset{\underset{\displaystyle OH}{|}}{P}}-OH \rightleftharpoons HO-\overset{\overset{\displaystyle O}{\|}}{\underset{\underset{\displaystyle OH}{|}}{P}}-O-\overset{\overset{\displaystyle O}{\|}}{\underset{\underset{\displaystyle OH}{|}}{P}}-OH + H_2O$$

pyrophosphoric acid

If ionized or if connected to an organic molecule, the ion or group is called a *pyrophosphate*. Pyrophosphoric acid can also form esters in the same way phosphoric acid does:

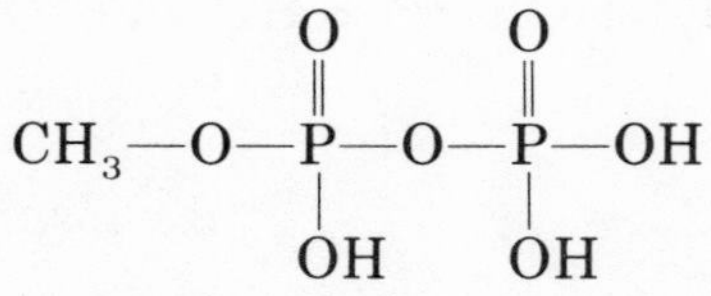

$$CH_3-O-\overset{\overset{\displaystyle O}{\|}}{\underset{\underset{\displaystyle OH}{|}}{P}}-O-\overset{\overset{\displaystyle O}{\|}}{\underset{\underset{\displaystyle OH}{|}}{P}}-OH$$

methyl pyrophosphate

Note that the same method is used to name pyrophosphate esters as phosphate esters except that *pyrophosphate* is substituted for *phosphate*. Some pyrophosphate esters are also important biologically. In general, phosphate groups can be attached to any other molecule that has an (—OH)-group, by eliminating a molecule of water.

The reverse of phosphate ester formation is phosphate ester hydrolysis, where one molecule of water is added to break the ester linkage:

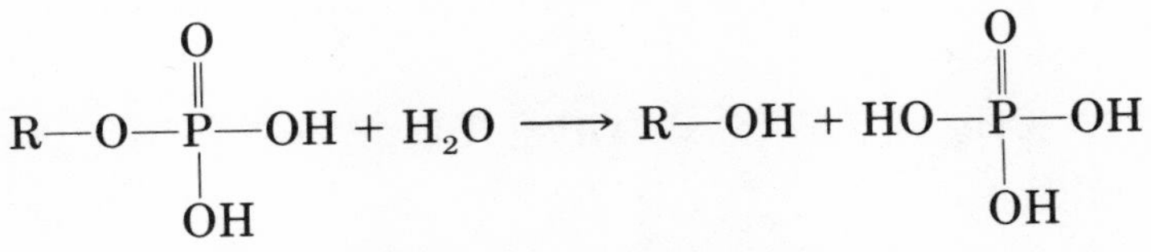

$$R{-}O{-}\overset{\overset{\displaystyle O}{\|}}{\underset{\underset{\displaystyle OH}{|}}{P}}{-}OH + H_2O \longrightarrow R{-}OH + HO{-}\overset{\overset{\displaystyle O}{\|}}{\underset{\underset{\displaystyle OH}{|}}{P}}{-}OH$$

hydrolysis of a phosphate ester

Analogous hydrolytic reactions (reactions that add water to break a bond) occur for all of the phosphate linkages described thus far.

Phosphate may form two esters and hence join two alcohol groups:

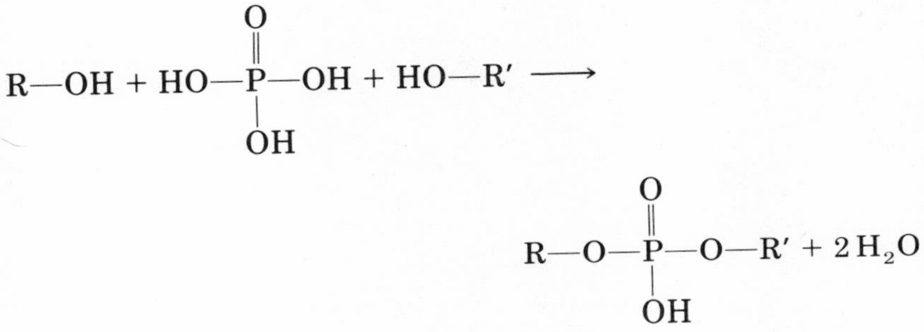

$$R{-}OH + HO{-}\overset{\overset{\displaystyle O}{\|}}{\underset{\underset{\displaystyle OH}{|}}{P}}{-}OH + HO{-}R' \longrightarrow$$

$$R{-}O{-}\overset{\overset{\displaystyle O}{\|}}{\underset{\underset{\displaystyle OH}{|}}{P}}{-}O{-}R' + 2\,H_2O$$

Such a *di*ester linkage is found in nucleic acids (DNA and RNA).

EXERCISES

1. Name the following compounds.

(a) 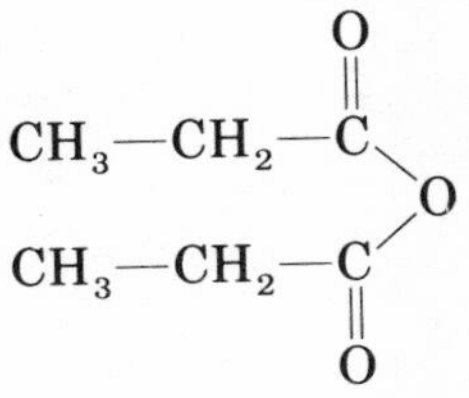 $CH_3{-}CH_2{-}\overset{\overset{\displaystyle O}{\|}}{C}{-}O{-}\underset{\underset{\displaystyle O}{\|}}{C}{-}CH_2{-}CH_3$

(b) $CH_3{-}CH_2{-}CH_2{-}SH$

(c) $CH_3{-}S{-}CH_3$

(d) 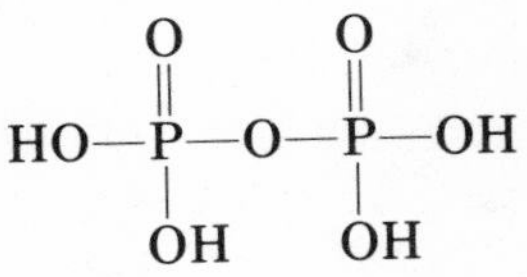 $HO{-}\overset{\overset{\displaystyle O}{\|}}{\underset{\underset{\displaystyle OH}{|}}{P}}{-}O{-}\overset{\overset{\displaystyle O}{\|}}{\underset{\underset{\displaystyle OH}{|}}{P}}{-}OH$

2. Show the reaction for the synthesis of propanamide from propanoic or propionic anhydride and ammonia.

3. Complete the following reaction showing the hydrolysis of a phosphate ester.

$$CH_3-CH_2-CH_2-O-\overset{\overset{\displaystyle O}{\|}}{\underset{\underset{\displaystyle OH}{|}}{P}}-OH + H_2O \longrightarrow$$

4. Draw the structures for the following compounds.

(a) The acetate derivative of methyl amine (the amide derived from acetic anhydride and methyl amine).

(b) The diacetate (ester) derivative of ethylene glycol ($HO-CH_2-CH_2-OH$).

(c) Heptyl pyrophosphate.

(d) The thioester that could be derived from acetic acid and n-butyl mercaptan.

Naming Multifunctional Compounds

As has already been shown, organic compounds may have more than one functional group. In fact, many biological compounds are polyfunctional (have several functional groups). One important class consists of amino acids:

$$CH_3—\underset{\displaystyle NH_2}{\underset{|}{CH}}—\overset{\displaystyle O}{\overset{\|}{C}}—OH$$

an amino acid called alanine (trivial), or
2-aminopropanoic acid (Geneva)

The presence of more than one functional group complicates naming compounds. In our previous discussions on naming, the appropriate suffixes for the different functional groups were given, but as indicated earlier we cannot use more than one suffix; therefore prefixes for the functional groups are also used. Table 8.1 indicates the names used as prefixes and suffixes for some functional groups.

Which group should be selected to be used as a suffix? Generally, the suffix is considered to be the most important classification of the compound. There is a formal priority system, but for our purposes the following outline is acceptable. If a carboxylic acid function or a carboxylic acid derivative (ester, amide, or anhydride) is present, it should be designated with the correct suffix. All other groups should be indicated by prefixes. The priority of the prefixes is generally in the order of listing in Table 8.1.

Recall that if there is more than one of the same group attached to a compound, a *di-*, *tri-*, or *tetra-* is added to the prefix. Thus, if two amino groups are present, the prefix becomes *diamino-* and the locations of these two amino groups are indicated by *a number for each.* The numbers immediately precede the prefix.

Several examples illustrating the naming procedure will now be considered.

1. $CH_3—CH(NH_2)—CH_2—CH_2—OH$

This compound could be named as an amine, meaning that the amine suffix is used, but let us name it as an alcohol. This specifies that the suffix be *-ol.* The longest continuous carbon chain is four carbon atoms; therefore the base name is *butane.* Putting the suffix on the base name yields *butanol.* The position

TABLE 8.1 ***Prefixes and suffixes for naming compounds with various functional groups.***

Functional Group	Prefix	Suffix
Alkane	-yl-*	-ane
C=C	-ene- (used in the middle of name; not a true prefix)	-ene
$—NH_2$	amino-	-amine
—OH	hydroxy-	-ol
—SH	mercapto-	-thiol
C=O	oxo-, keto- (sometimes)	-one
—C(=O)—H	formyl-*	-al
—C(=O)—OH	carboxy-*	-oic acid
R—C(=O)—	-yl-*	—
$—C(=O)—NH_2$		-amide
—C(=O)—OR′		-oate with name of R′ preceding the acid name

* The carbon atom is not counted in the chain that serves as the compound's base name.

of the —OH must be indicated. Since the lowest numbers possible should be used, the carbon atoms are numbered from the —OH-group. So the name becomes *1-butanol.* All that remains is to indicate that there is an amino group attached to the third carbon atom. The complete name should be *3-amino-1-butanol.* The molecule looks exactly like 1-butanol with an amino group, which the name clearly indicates.

2. 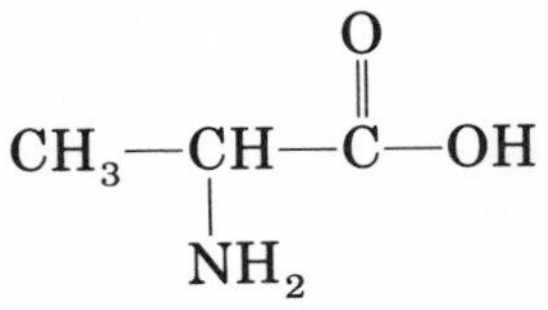 $$CH_3\text{—}\underset{\underset{NH_2}{|}}{CH}\text{—}\overset{\overset{O}{\|}}{C}\text{—}OH$$

2-aminopropanoic acid
(alanine)

Name dissection: There is an acid group and an amino group. We should choose to name this as an acid (use the acid system as the suffix). There are three carbon atoms, and so the base name is *propane*. Changing to the correct suffix for an acid gives *propanoic acid*. An amino group is attached to the #2 carbon atom. Therefore the prefix *2-amino-* should be attached to the acid's name to yield *2-aminopropanoic acid*.

What should we do if an alkyl radical is attached to an amino group but it is not desirable to name the compound as an amine? As indicated for amides, a capital *N* preceding a radical means that the radical is attached to a nitrogen atom.

3. 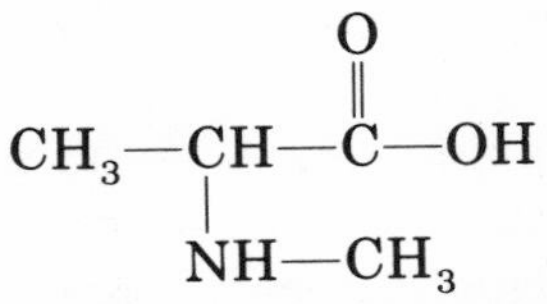 $$CH_3\text{—}\underset{\underset{NH\text{—}CH_3}{|}}{CH}\text{—}\overset{\overset{O}{\|}}{C}\text{—}OH$$

N-methyl-2-aminopropanoic acid

Name dissection: In this case the 2-amino group has an *N*-methyl group attached to it. The rest of the name should be obvious from the previous example.

4. 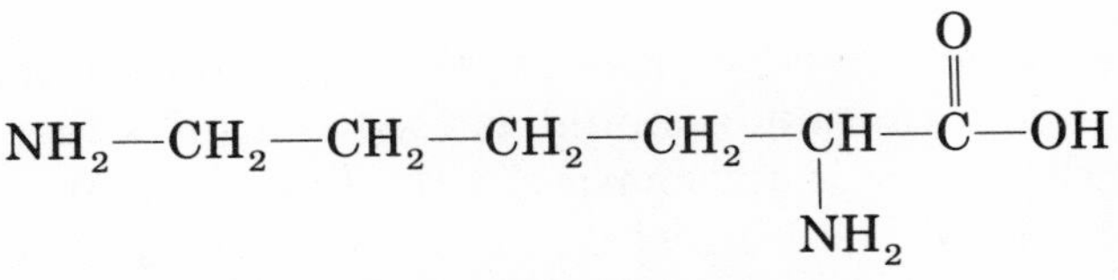 $$NH_2\text{—}CH_2\text{—}CH_2\text{—}CH_2\text{—}CH_2\text{—}\underset{\underset{NH_2}{|}}{CH}\text{—}\overset{\overset{O}{\|}}{C}\text{—}OH$$

2,6-diaminohexanoic acid
(lysine)

Name dissection: First, the carbon skeleton shows that there are six carbon atoms in a straight chain; hence the base name should be *hexane*. Second, there are two amino groups and one acid group. Therefore this compound should be named as an acid, and the carboxyl group of the acid group becomes the #1 carbon atom. The name should end in *-hexanoic acid*. Third, we need to indicate that there are two amino groups present and specify where they are located. The correct prefix should be *diamino-* and the locations are the #2 and #6 carbon atoms.

5. $CH_3—CH(OH)—CH(NH_2)—C(=O)—O—CH_3$

methyl 2-amino-3-hydroxybutanoate
(methyl ester of threonine)

Name dissection: First, there are four carbon atoms in a straight chain; hence the base name is *butane*. Second, there is an ester group, and so the name should consist of two words. The first word should be the name of the radical representing the alcohol portion. This name must always be a separate word and be the very first part of the name. The ending of the second word must be *-oate*. Therefore we should have *Methyl____ butanoate*. Third, there is an amino group at carbon atom #2 and a hydroxyl group at carbon atom #3.

Do the prefixes have to be given in a particular order? The rule is that we can give the groups either in alphabetical order or in order of decreasing complexity. Because it is not always easy to determine relative complexity of groups, alphabetical order is used more frequently. The part of the prefix indicating the number of groups is ignored; for example, *dimethyl* is treated alphabetically as *methyl*, and *trihydroxy* as *hydroxy*.

Because the common name for 2-amino-3-hydroxybutanoic acid is *threonine*, calling this compound the *methyl ester of threonine* is perfectly descriptive and acceptable.

6. $CH_3—CH(CH_3)—CH(NH_2)—C(=O)—NH_2$

Amide, amino, and methyl groups are present. Therefore the choice is easy; the amide should be used as a suffix, and the carbonyl carbon atom is automatically the #1 carbon in acids, amides, and esters. The longest continuous carbon chain has four atoms; hence the base name is *butane*. Combining the base name and the appropriate suffix gives *butanamide*. Next, the positions of the other groups should be indicated. There is a methyl group on the #3 carbon atom and an amino group on the #2 carbon atom; thus the full name should be *2-amino-3-methylbutanamide*. The following is an even more complex structure.

$$\begin{array}{c} CH_3\ \ OO \\ 7.\ \ CH_3-\underset{OH}{CH}-\underset{OH}{CH}-CH-\overset{\|}{C}-CH_2-CH_2-\overset{\|}{C}-OH \end{array}$$

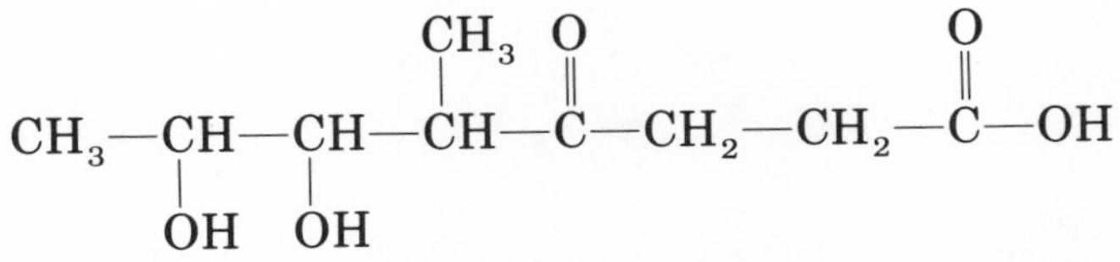

Here, there are two hydroxyl groups, one keto group (ketone), one methyl group, and one acid group. We should choose to name this as an acid. There are eight carbon atoms in the longest chain; therefore *octane* is the base name. As an acid this would be *octanoic acid.* Next, the kinds of groups attached and their positions must be indicated. Remember, since the suffix designates an acid, the carboxyl carbon atom is the #1 carbon atom. There are two hydroxyl groups on carbon atoms #6 and #7, a methyl group on carbon #5, and a keto group on carbon #4. For the keto group the prefix *oxo-* is used. Therefore the name should be *6,7-dihydroxy-5-methyl-4-oxooctanoic acid.* Again the alphabetical order has been used for listing substituent groups.

Up to this point we have used the Geneva system to name multifunctional compounds. The trivial system is similar, but Greek letters instead of numbers are used to indicate the positions of the various groups. This is best illustrated in acids or acid derivatives:

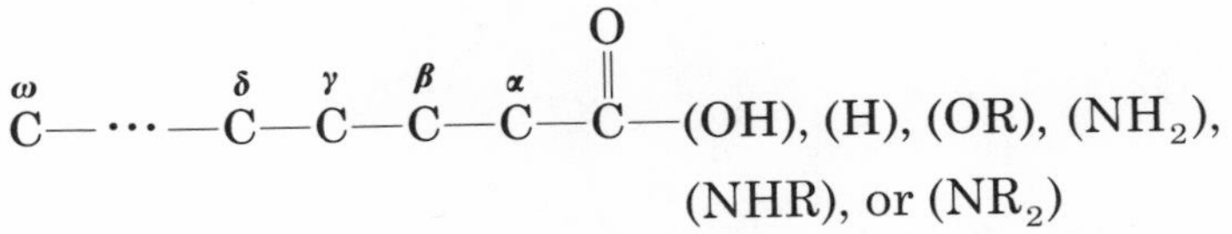

$$\overset{\omega}{C}-\cdots-\overset{\delta}{C}-\overset{\gamma}{C}-\overset{\beta}{C}-\overset{\alpha}{C}-\overset{O}{\overset{\|}{C}}-(OH), (H), (OR), (NH_2), (NHR), \text{ or } (NR_2)$$

Unlike in the Geneva system, the carbon atom of the *carboxyl group is not counted.* Instead, the first carbon atom *following* the carbonyl is labeled α, the second β, the third γ, the fourth δ, and so on through the Greek alphabet. One special point should be remembered: If an ω *is used, it always indicates the last carbon atom* in the chain, regardless of how long the chain is.

$$8.\ \ H_2N-CH_2-CH_2-CH_2-\overset{O}{\overset{\|}{C}}-OH$$

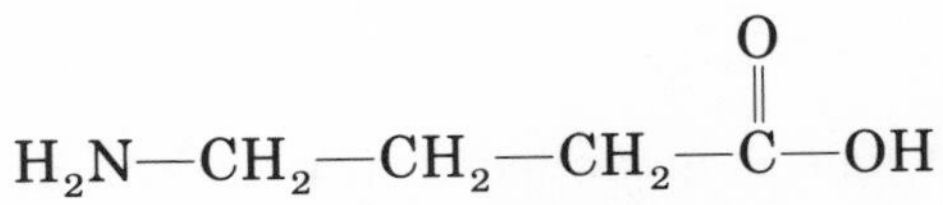

γ-aminobutyric acid

Name dissection: This should be named as an acid rather than as an amine, and there are four carbon atoms. The trivial name of the four-carbon monocarboxylic acid is *butyric acid.* Hence that is the correct acid name to use, not *butanoic*, which is used in the Geneva system.

Because there is one amino group attached to the γ-carbon atom, that is added as a prefix to the acid's name. While ω is generally used for long chains, we could, but probably would not, name the compound *ω-aminobutyric acid.*

9. 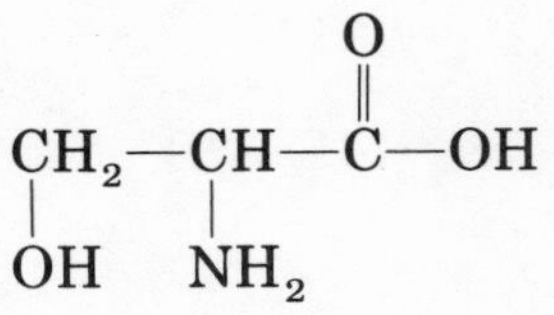 $\underset{\displaystyle OH}{\underset{|}{CH_2}}-\underset{\displaystyle NH_2}{\underset{|}{CH}}-\overset{\displaystyle O}{\overset{\|}{C}}-OH$

serine

Serine is the common name for this amino acid but it could be named systematically using the trivial system. In this case, it should be named as an acid. There are three carbon atoms and the correct acid name in the trivial system is *propionic acid.* There is an amino group on the α-carbon and a hydroxyl group on the β-carbon. Therefore the name should be *α-amino-β-hydroxypropionic acid.* By the Geneva system it would be *2-amino-3-hydroxypropanoic acid.* From these names we can see why the name serine is frequently preferred to the other two alternatives.

Extensive rules have been developed for naming compounds, but much more space would be necessary to explore all of the fine points. To the author, a name that describes a compound in unambiguous terms is acceptable since that is the purpose of a name. Nevertheless, we should not deliberately mix the Geneva and trivial systems. For complex molecules the Geneva system generally is easier to apply.

8.1 SOME SPECIAL NAMES

8.1.1 Carbonyl Groups

The carbonyl group, briefly defined previously, is found in several functional groups. The common feature is

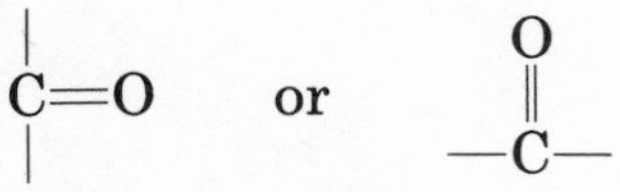

$$\underset{|}{\overset{|}{C}}=O \quad \text{or} \quad -\overset{\displaystyle O}{\overset{\|}{C}}-$$

It does not matter what else is attached to the carbon atom. Therefore aldehydes, ketones, carboxylic acid, acid anhydrides, esters, and amides all contain a carbonyl group.

8.1.2 Acyl Groups

In Chapter 5 acyl groups ($R-\overset{\displaystyle O}{\overset{\|}{C}}-$) were discussed in connection with their presence in esters. However, acyl groups are present in all of the various compounds that may be considered to be carboxylic acid derivatives, such as esters, amides, and anhydrides.

8.1.3 Aliphatic Compounds

Organic compounds may be divided into two broad categories: aliphatic compounds and aromatic compounds. The easiest way to define aliphatic compounds is to state that they are not aromatic compounds. The distinctive feature of aromatic compounds will be given in Chapter 10. So far, only aliphatic compounds have been discussed, and in general they contain open carbon chains or rings *without* alternating double and single bonds.

The following exercises provide practice in naming compounds. Additional exercises for naming are given in Section A.3 of the Appendix.

EXERCISES

1. Name the following compounds by the *Geneva* system.

(a) $CH_3-CH_2-CH(NH_2)-CH_2-C(=O)-OH$

(b) $HO-CH_2-CH_2-CH_2-C(=O)-H$

(c) $CH_3-CH_2-CH_2-C(=O)-O-CH_3$

(d) $CH_3(CH_2)_4CH(CH_3)-C(=O)-NHCH_3$

(e) $CH_3-CH_2-CH(CH_3)-C(CH_3)_2-CH(NH_2)-C(=O)-NH_2$

(f) $CH_3-CH_2-CH(CH_3)-CH(OH)-C(=O)-O-CH_2-CH_3$

2. Name the following compounds by the trivial system.

(a) 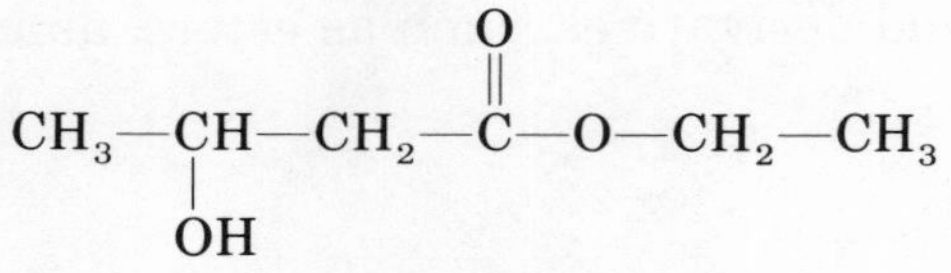 $$CH_3-\underset{\displaystyle OH}{\underset{|}{CH}}-CH_2-\overset{\displaystyle O}{\overset{\|}{C}}-O-CH_2-CH_3$$

(b) 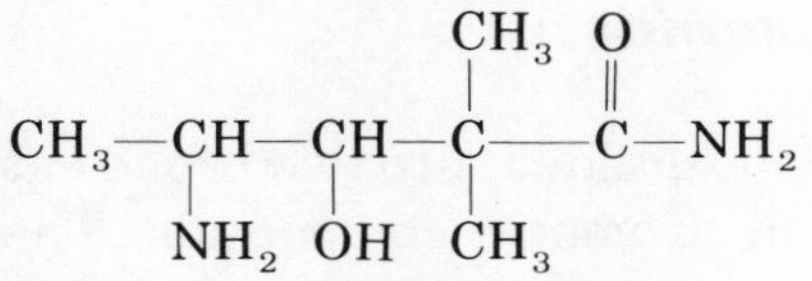 $$CH_3-\underset{\displaystyle NH_2}{\underset{|}{CH}}-\underset{\displaystyle OH}{\underset{|}{CH}}-\overset{\displaystyle CH_3}{\overset{|}{\underset{\displaystyle CH_3}{\underset{|}{C}}}}-\overset{\displaystyle O}{\overset{\|}{C}}-NH_2$$

3. Write the structure for the following compounds.
(a) methyl 2-amino-3-hydroxypentanoate
(b) 6-hydroxy-3,3-dimethyl-4-oxoheptanoic acid
(c) N-ethyl-3-hydroxy-2-methylbutanamide
(d) γ-amino-α-hydroxy-β-methylbutyric acid
(e) 2-N-methylamino-4-oxo-1-hexanol
(f) propanoyl phosphate

Optical Isomers

A carbon atom that has four *different* atoms or groups attached to it is called an *asymmetric* or *chiral* carbon atom. The following two molecules contain an asymmetric carbon atom, but, as written, bond rotation would seem to rule out any possible differences between the molecules. However, models will show that indeed they are different.

```
      H           H
      |           |
  Br—C—I       I—C—Br
      |           |
      F           F
```

Figure 9.1 contains two models of the molecules shown here. They are mirror images of each other, just like our right and left hands. Can we superimpose one on the other? To be superimposed, the H must be oriented upward; thus rotation about the axis of the H–C bond is the only hope to achieve the same orientation. Rotating the model on the right two-thirds of a turn would allow H, C, and Br to be aligned, but F and I would be reversed. Further rotation will allow three atoms to be aligned but the other two are *always* reversed. The conclusion is that these models are not superimposable. According to our previous definition, isomers are compounds that have the same molecular formula but whose molecular models cannot be superimposed. Therefore the models shown as mirror images in Fig. 9.1 represent isomers. Such *pairs* (mirror images) are called *enantiomers* or *enantiomorphs.*

FIGURE 9.1

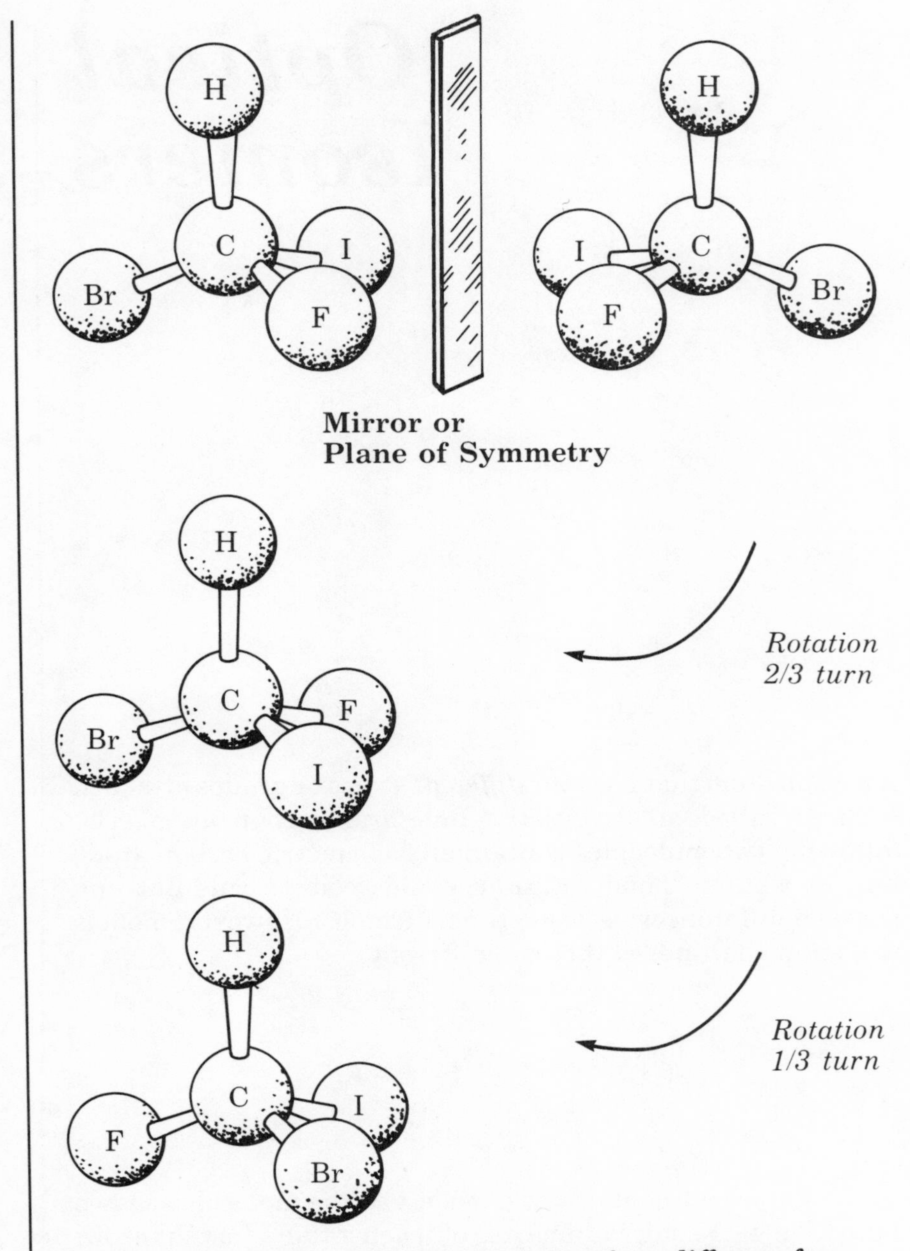

Asymmetry of a carbon atom bearing four different functional groups. The two models shown at the top appear as mirror images. Attempts to align the atoms of the right model with those of the model on the left will be unsuccessful. To maintain the alignment of the C and H atoms, the model on the right can be rotated only about the axis of the C—H bond. Rotation two-thirds of a turn (clockwise) permits the alignment of the C, H, and Br atoms, but the F and I atoms are reversed. Rotation one-third of a turn permits alignment of the C, H, and I atoms, but the F and Br atoms are reversed. The conclusion is that the upper-right model cannot be oriented so that its atoms may be superimposed upon the atoms of the left model. Hence the upper models represent isomers—in this case, optical isomers.

One should not assume that all pairs of mirror images represent enantiomers. Figure 9.2 shows two models that are mirror images. However, rotation of the model on the right one-half of a turn will allow the models to be superimposed. Therefore they *do not represent isomers*; but the molecule *does not contain an asymmetric carbon atom either*, that is, the carbon atom does not bear four different groups. As shown in Fig. 9.3 a plane of symmetry exists in $CHFI_2$; that is, if a straight slice through the centers of the C, H, and F atoms were made, the two halves of the molecule would be mirror images. There is no way that CHBrFI could be sliced to get halves of the molecule that are mirror images; therefore the molecule itself is asymmetric. Optical isomers are *not symmetrical* molecules, which explains the reason for the term *asymmetrical*. Remember that symmetry within a molecule eliminates the possibility of optical isomers and that a chiral carbon atom

FIGURE 9.2

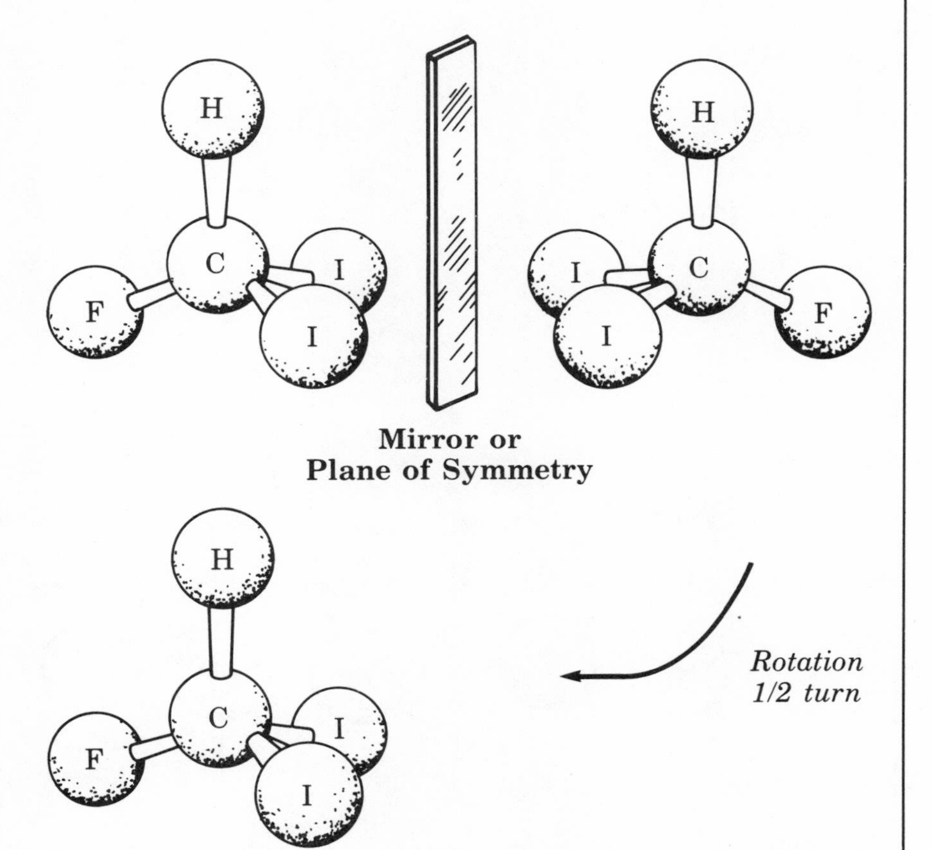

Mirror images of models that do not contain an asymmetrical carbon atom. The atoms of the upper-right model can be aligned with the atoms of the upper-left model by clockwise rotation of the model. This allows the models to be superimposable and hence the models represent the same compound, not isomers.

(one with four different groups) introduces asymmetry into a molecule. Although enantiomers are always mirror images of one another, mirror images are not always enantiomers or even optical isomers.

To determine if optical isomers are possible for a given structure, always search for chiral or asymmetric carbon atoms. If *asymmetric carbon atoms* are *present* you can be confident that *optical isomers exist*. Actually, there are a few cases where another type of symmetry may be introduced into a molecule, which makes it possible to have asymmetric carbon atoms without the compound existing in different isomeric forms. Such cases will not be considered here.

If you review the structures presented in previous chapters you will find that many contain asymmetric carbon atoms, a feature that was simply ignored. For practice, determine which of the following four compounds contain an asymmetric carbon atom:

1. CH_3—CH—NH_2 (with OH on the central CH)

2. H—CH—CH_2—CH_3 (with NH_2 on the CH)

3. CH_3—CH—CH_2—CH_3 (with OH on the CH)

FIGURE 9.3

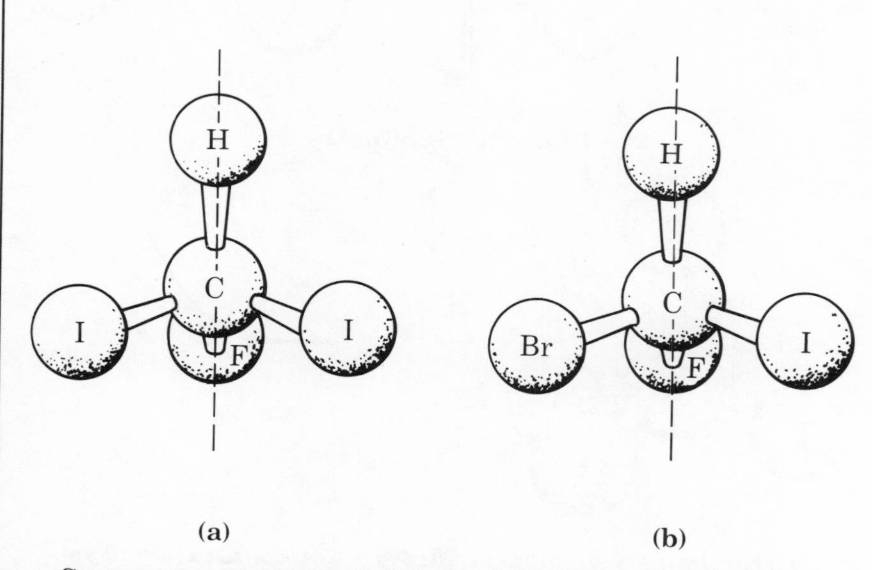

Symmetry within a molecule. A plane of symmetry exists in the molecule represented by (a) but not in that represented by (b). The molecule represented by (b) has an asymmetric carbon atom whereas (a) does not. The center line represents the edge of a plane that extends backward.

$$
\begin{array}{c}
\text{OH} \\
| \\
\textbf{4.}\ \text{CH}_3\text{—C—CH}_3 \\
| \\
\text{NH}_2
\end{array}
$$

Compound (1) does. The four different groups are —H, —OH, —NH_2, and —CH_3. Compound (3), 2-butanol, has an asymmetric carbon, bearing —H, —OH, —CH_3, and —CH_2—CH_3. Note that the whole group attached to the carbon atom is compared. In this case the asymmetric carbon atom is attached to *two* other carbon atoms, but those carbon atoms are parts of different groups—methyl and ethyl groups. Because of two identical groups, (H—) in compound (2) and (CH_3—) in compound (4), those compounds do not contain an asymmetric carbon atom.

Compounds that contain asymmetric carbon atoms are said to be *optically active* because they exhibit the striking ability to rotate a plane of polarized light. Enantiomers, which are mirror images of each other, rotate such light in opposite directions! Polarimeters are optical devices used to measure rotation of polarized light caused by optically active compounds. Equal quantities of two enantiomers in a mixture will not rotate a plane of polarized light because they cancel each other's effect. One enantiomer may rotate the plane clockwise, and it is said to be *dextrorotatory*. The other enantiomer would rotate the plane by the same amount, but in the counterclockwise direction, and that enantiomer would be *levorotatory*. A (+) in front of a name indicates that the dextrorotatory enantiomer is the one being named and a (−) stands for the levorotatory enantiomer. Unlike with all other kinds of isomers, the usual physical properties of enantiomers (melting point, boiling point) are identical; only their abilities to rotate polarized light differ.

The asymmetry of optical isomers is important to living organisms. The majority of the compounds that make up an organism are optically active and essentially all chemical reactions that occur in life processes involve only one enantiomorph.

Two systems are commonly used to distinguish between enantiomorphs. In the *Cahn, Ingold, and Prelog method*, one enantiomer is designated as *R* and the other as *S*. The rotational order of groups according to a priority is determined and if that direction is clockwise the configuration is designated as *R*, which stands for the Latin word *rectus*, meaning "right." If the direction of the priority order is counterclockwise an *S* is used, which stands for the Latin word *sinister*, meaning "left." Although this system is increasing in popularity, the other system is more prevalent in biochemistry.

The *D and L system* is based on the configuration of the two enantiomers of glyceraldehyde:

$$\begin{array}{l} \text{O} \\ \| \\ \text{C—H} \\ | \\ \text{CH—OH} \\ | \\ \text{CH}_2\text{—OH} \end{array}$$

glyceraldehyde

The middle carbon atom of glyceraldehyde is asymmetric. It has an $—H_2$, —OH, —CHO, and CH_2—OH-group attached to it:

$$\begin{array}{ccc} \text{CHO} & \qquad & \text{CHO} \\ \text{H}\blacktriangleright\text{C}\blacktriangleleft\text{OH} & & \text{HO}\blacktriangleright\text{C}\blacktriangleleft\text{H} \\ \text{CH}_2\text{OH} & & \text{CH}_2\text{OH} \end{array}$$

D-glyceraldehyde L-glyceraldehyde

The two forms of glyceraldehyde are given here with the asymmetric carbon atom shown in the plane of the paper. The dashed bonds indicate that the aldehyde group and the CH_2—OH-group project *behind* the plane of the paper. Both the H- and OH-groups project *in front of* the plane of the paper. If the OH-group is on the right side of the asymmetric carbon it

FIGURE 9.4

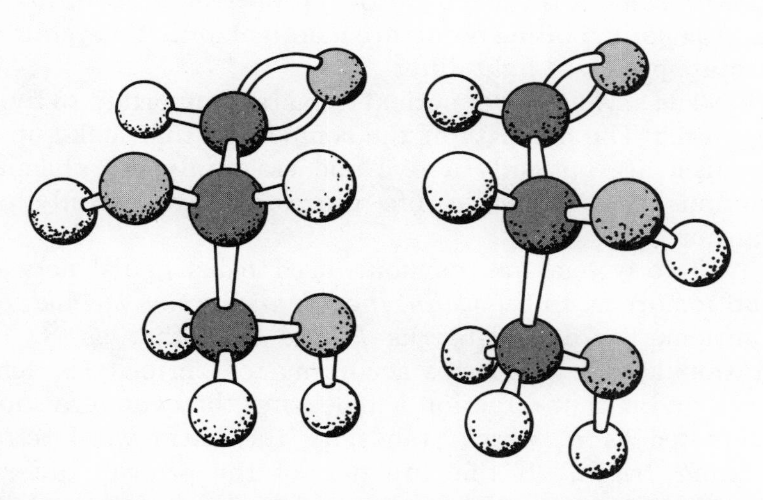

Ball-and-stick models of D- and L-glyceraldehyde, which are the reference compounds for designating D- and L-isomers of other compounds. The D-form is on the right, and the L-form on the left.

is designated as the D-form, whereas if it is on the left side it is designated as the L-form. Figure 9.4 shows models of the two enantiomers of glyceraldehyde to help visualize the projection structures.

To simplify writing structures and avoid using dashed or flared lines, *Fischer projections* are used. Here, the carbon chain is written vertically rather than horizontally and the —CHO-group is oriented upward. The —OH- and —H-groups are written on the sides where they would project out from the plane of the paper, and the other group (—CH_2OH in this case) is oriented downward. A modified appearance results, but the placement of the groups in space remains identical to that shown by the projection structures. Fischer projections are as follows:

$$\begin{array}{c} CHO \\ | \\ H—C—OH \\ | \\ CH_2OH \end{array} \qquad \begin{array}{c} CHO \\ | \\ HO—C—H \\ | \\ CH_2OH \end{array}$$

D-glyceraldehyde   L-glyceraldehyde

Consider an acid with an asymmetric carbon atom such as lactic acid,

$$\begin{array}{c} CH_3—CH—CO_2H \\ | \\ OH \end{array}$$

For glyceraldehyde it was stated that the —CHO should be oriented upward. A more general rule is followed to permit asymmetric molecules other than aldehydes to be written as Fischer projections. The most highly oxidized portion of the molecule is oriented upward. In lactic acid the carboxyl carbon atom is more highly oxidized than the carbon atom bearing the OH-group or the carbon atom of the methyl group. Hence the carboxyl group should be oriented upward. The —OH- and —H-groups are written on the sides (right or left) where they would project outward from the plane of the page. Again, if the OH-group is on the left side of the asymmetric atom, that isomer is designated as the L-isomer; if on the right, it is the D-isomer:

$$\begin{array}{c} CO_2H \\ | \\ H—C—OH \\ | \\ CH_3 \end{array} \qquad \begin{array}{c} CO_2H \\ | \\ HO—C—H \\ | \\ CH_3 \end{array}$$

D-lactic acid   L-lactic acid

Amino acids may be designated similarly:

$$\begin{array}{ccc} & CO_2H & \\ & | & \\ H- & C & -NH_2 \\ & | & \\ & CH_3 & \end{array} \qquad \begin{array}{ccc} & CO_2H & \\ & | & \\ H_2N- & C & -H \\ & | & \\ & CH_3 & \end{array}$$

D-alanine L-alanine

The carbon atom bearing the amino group, like a hydroxyl group, is less highly oxidized than the carbon atom in the carboxyl group, which dictates that the carboxyl group be the uppermost group in the vertical chain. Those compounds that have the same configuration about the asymmetric carbon as D-glyceraldehyde belong to the class called the D-series. The L-series comprises compounds that have the same configuration as L-glyceraldehyde. Thus the D-series has a functional group such as an OH— or NH_2— to the *right* and the L-series has them to the *left* when the most highly oxidized group is oriented upward. The orientation of the hydrogen atom about the chiral carbon atom is the same for all members of the D-series and the same is true for the members of the L-series.

9.1 DIASTEREOISOMERS

What if there are two asymmetric carbon atoms in the same molecule? For example, consider the following:

$$\begin{array}{ccc} & CHO & \\ & | & \\ H- & C & -OH \\ & | & \\ H- & C & -OH \\ & | & \\ & CH_2OH & \end{array} \quad \begin{array}{ccc} & CHO & \\ & | & \\ HO- & C & -H \\ & | & \\ H- & C & -OH \\ & | & \\ & CH_2OH & \end{array} \quad \begin{array}{ccc} & CHO & \\ & | & \\ H- & C & -OH \\ & | & \\ HO- & C & -H \\ & | & \\ & CH_2OH & \end{array} \quad \begin{array}{ccc} & CHO & \\ & | & \\ HO- & C & -H \\ & | & \\ HO- & C & -H \\ & | & \\ & CH_2OH & \end{array}$$

D-erythrose D-threose L-threose L-erythrose

In each of these four optical isomers there are two asymmetric carbon atoms. The same vertical orientation described for glyceraldehyde is used in this case, but the D-isomers have *the hydroxyl group furthest from the aldehyde group* on the right, whereas the L-isomers have the hydroxyl group furthest from the aldehyde group on the left. The D-series consists of two members that have opposite configurations about the remaining asymmetric carbon atom. To distinguish between these two isomers, a completely different name is used. *D-erythrose* and *D-threose* are *diastereoisomers*, not *enantiomers*. *D-erythrose* and *L-erythrose* have *opposite* configurations at each asymmetric carbon atom; therefore they are *enantiomers* and their models are *mirror images.* Likewise, D-threose and L-threose are enantiomers and L-threose and L-erythrose are diastereoisomers. These four isomers are members of the family of simple sugars, or monosaccharides, which are important in biochemistry. The presence of a third asymmetric carbon atom complicates the matter further, but the same system is extended to designate the various isomers.

The general procedure for designating optical isomers is summarized diagrammatically as follows:

Structure		Note
CHO	⟵	*most highly oxidized group at the top*
¦	⟵	*vertical arrangement of carbon chain*
H—C—OH	⟵	*configuration about the asymmetric carbon atom furthest down from the most highly oxidized carbon atom determines if the isomer is a member of the D-series (as shown here) or the L-series*
¦		
CH_2OH		

A carbon chain with these features is directly comparable to glyceraldehyde. Replacement of the dashed lines (portions of the carbon chain) by bonds yields glyceraldehyde.

Enantiomers have the same word for a name but are distinguished by a D- or an L-prefix, and they are always mirror images of one another. Sometimes the D- or L-designations are omitted, as in the case of *ribose.* The reader cannot be sure which isomer or if a D and L mixture of isomers is meant. There is some justification for the shortcut in biochemistry because almost always the D-isomer (D-ribose) is encountered. Therefore one would assume that *ribose* meant D-*ribose.* Had L-*ribose* been intended it would have been written that way since it is an uncommon form. Of course, knowledge about the isomers present in biological systems must be gained before the shortcut is justified.

Consider the eight optical isomers of five-carbon sugars, for example:

D-ribose	D-arabinose	D-xylose	D-lyxose
CHO	CHO	CHO	CHO
H—C—OH	HO—C—H	H—C—OH	HO—C—H
H—C—OH	H—C—OH	HO—C—H	HO—C—H
H—C—OH	H—C—OH	H—C—OH	H—C—OH
CH_2OH	CH_2OH	CH_2OH	CH_2OH

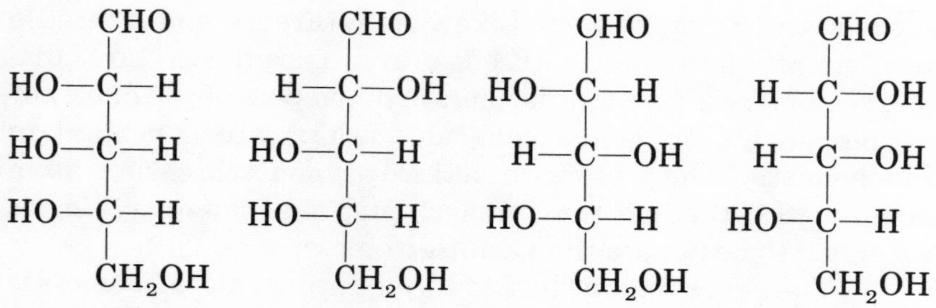

```
    CHO               CHO                CHO               CHO
     |                 |                  |                 |
HO—C—H            H—C—OH            HO—C—H            H—C—OH
     |                 |                  |                 |
HO—C—H           HO—C—H              H—C—OH            H—C—OH
     |                 |                  |                 |
HO—C—H           HO—C—H             HO—C—H            HO—C—H
     |                 |                  |                 |
    CH2OH             CH2OH              CH2OH             CH2OH
```

L-ribose L-arabinose L-xylose L-lyxose

There are six diastereoisomers of D-ribose, which are D- and L-arabinose, D- and L-xylose, and D- and L-lyxose. D-ribose has only one enantiomer, L-ribose. The configuration of each asymmetric carbon atom in L-ribose is the reverse of that in D-ribose, and like other enantiomers they are mirror images. Analogous relationships exist for the other five-carbon sugars.

By reviewing the number of isomers for the simple sugars we see that the number of optical isomers of a compound is related to the number of chiral or asymmetric carbon atoms present. The mathematical relationship is

$$\text{Number of isomers} = 2^n$$

where n = number of asymmetric carbon atoms. Hence for one chiral atom, $2^n = 2^1 = 2$ optical isomers are possible. Glyceraldehyde has one asymmetric carbon atom and two isomers. For the five-carbon sugars there are three asymmetric atoms and $2^3 = 8$ optical isomers. The presence of several asymmetric carbon atoms in a biological compound is common, but more often than not only one isomer of such a compound will be present in a living organism.

EXERCISES

1. Place an asterisk above the asymmetric or chiral carbon atoms in the following compounds.

(a) 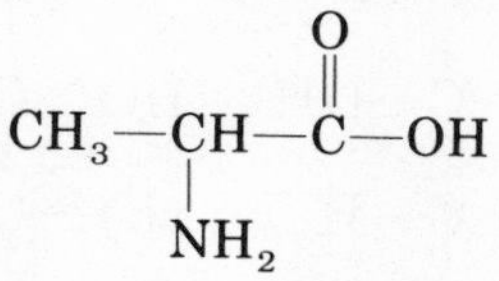
```
              O
              ‖
CH3—CH—C—OH
       |
      NH2
```

(b) 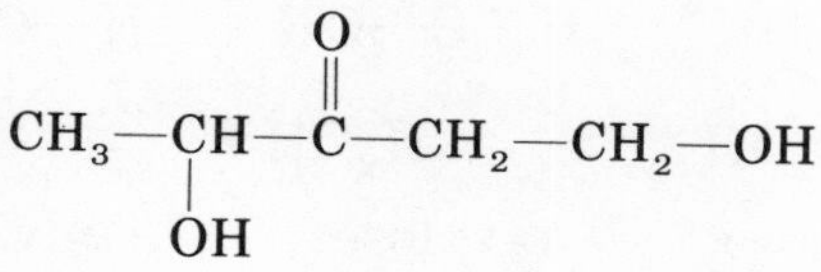
```
              O
              ‖
CH3—CH—C—CH2—CH2—OH
       |
      OH
```

(c)

$$\begin{array}{c} CH_3 \\ | \\ CH_3{-}CH_2{-}C{-}CH_2{-}CH_3 \\ | \\ CH_2{-}CH_3 \end{array}$$

(d)

$$\begin{array}{c} CH_3 \\ | \\ CH_3{-}CH_2{-}C{-}CH_2{-}OH \\ | \\ H \end{array}$$

(e)

(f) $CH_2{-}OH$

2. Indicate whether the following stereoisomers are of the D- or L-series by placing either D or L in the name. Compounds (a) through (c) are shown as Fischer projections, whereas (d) is a projection structure (see Section 1.2 and Fig. 1.1).

(a)

$$\begin{array}{c} CO_2H \\ | \\ H{-}C{-}NH_2 \\ | \\ CH_2{-}OH \end{array}$$

-serine

(b)

$$\begin{array}{c} CO_2H \\ | \\ H_2N{-}C{-}H \\ | \\ CH_2 \\ | \\ CH{-}CH_3 \\ | \\ CH_3 \end{array}$$

-leucine

(c)

$$\begin{array}{c} CHO \\ | \\ H{-}C{-}OH \\ | \\ CH_3 \end{array}$$

-2-hydroxypropanal

(d)

$$\begin{array}{c} CH_2OH \\ | \\ C \\ H \quad\quad OH \\ CHO \end{array}$$

-glyceraldehyde (*Hint*: You may need to rewrite this structure.)

3. Draw the enantiomer of each of the following compounds and label each D or L. The compounds are shown as Fischer projections.

(a)

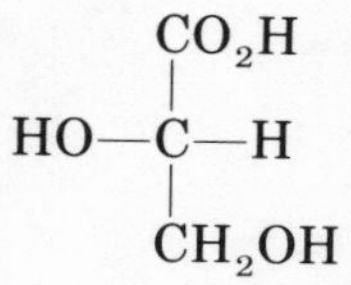

$$\begin{array}{c} CO_2H \\ | \\ HO—C—H \\ | \\ CH_2OH \end{array}$$

(b)

$$\begin{array}{c} CO_2H \\ | \\ HO—C—H \\ | \\ H—C—OH \\ | \\ CH_3 \end{array}$$

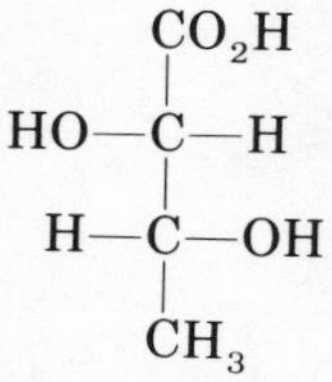

4. Draw at least one diastereoisomer of each of the following compounds and label each D or L. The compounds are shown as Fischer projections.

(a)

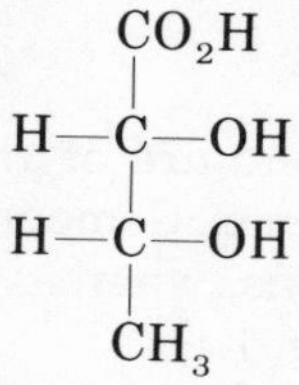

$$\begin{array}{c} CO_2H \\ | \\ H—C—OH \\ | \\ H—C—OH \\ | \\ CH_3 \end{array}$$

(b)

$$\begin{array}{c} CO_2H \\ | \\ H—C—OH \\ | \\ HO—C—H \\ | \\ H—C—OH \\ | \\ CH_2OH \end{array}$$

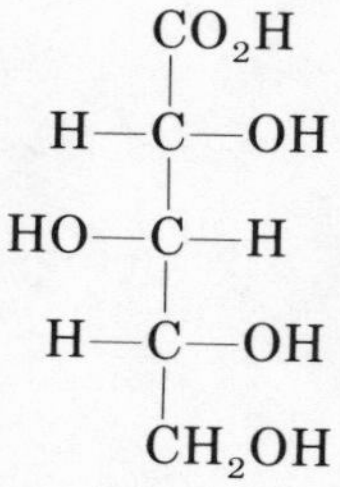

Aromatic Compounds

Aromatic compounds contain rings and multiple double bonds. They are often obtained from coal tar, which is produced by heating coal at a high temperature in the absence of air. Benzene (C_6H_6) is the simplest member of this large group of compounds:

```
         H                    H
         ··                   |
         C                    C
     ·        ·           /      \\
H:C          C:H     H—C          C—H
  ····        ··        ||         |
H:C          C:H     H—C          C—H
     ·        ·           \      //
         C                    C
         ··                   |
         H                    H
```

different representations of benzene

In the structural formula on the right it is *understood* that a carbon atom is located at each vertex of the hexagon and that the hydrogen atoms needed to satisfy the valences of the carbon atoms are attached. Note that in benzene the bonds between the carbon atoms are alternately double and single bonds. Double bonds separated by one single bond are said to be *conjugated*. Thus the double bonds are *fully conjugated*, meaning that double and single bonds alternate throughout the entire ring. Now aromatic compounds can be defined more precisely: They are compounds that have fully conjugated double bonds within a ring or rings. Consider the compounds

$CH_2{=}CH{-}CH{=}CH{-}CH{=}CH_2$ 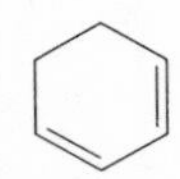

1,3,5-hexatriene 1,3-cyclohexadiene

1,3,5-Hexatriene has fully conjugated double bonds, but a ring is not present; thus it *is not* an aromatic compound. 1,3-Cyclohexadiene has a six-membered ring and the double bonds are conjugated, but the ring is not *fully* conjugated since alternation of double and single bonds does not extend throughout the whole ring. Therefore 1,3-cyclohexadiene is an *aliphatic*, rather than an aromatic, compound. On the surface, the classification of compounds as aromatic or aliphatic seems quite arbitrary. However, the distinction between such classes was originally made on the basis of physical and chemical properties rather than structural features. Subsequently, it was discovered that the fully conjugated ring(s) were responsible for the distinctive properties of aromatic compounds.

The double bonds in benzene simply do not behave like double bonds in aliphatic compounds. Kekulé was the first to explain the reason for this. Consider the two *resonance forms* of benzene:

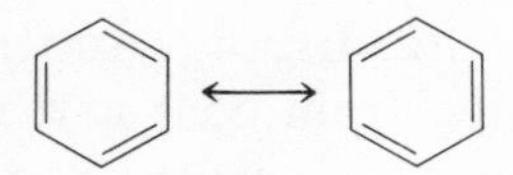

Kekulé proposed that the two resonance forms are in a dynamic state of oscillation; that is, a benzene molecule is always shifting back and forth between these two forms. If a single carbon atom could be observed we would expect to see that half of the time it would have a double bond with the carbon atom ahead of it clockwise, and half of the time it would have a double bond with the carbon atom behind it. *On the average* there would be $1\frac{1}{2}$ bonds between adjacent carbon atoms. In fact, each bond in benzene has partial single-bond and partial double-bond characteristics. Instead of continually shifting electrons back and forth, the actual structure is a *compromise*, or *hybrid*. Thus benzene might look like this in a "freeze-frame":

```
          H
          ..
          C
       .·   ·.
H:C           C:H
  ...         ...
H:C           C:H
       ·.   .·
          C
          ..
          H
```

or (dashed-bond benzene ring) *(Dashes indicate half-bonds.)*

Without a doubt, the bonds in benzene are not the same as those in aliphatic ring compounds such as cyclohexane.

Because of the bond angles of carbon (tetrahedron) a six-membered ring such as cyclohexane cannot exist as a flat ring. As shown in Fig. 2.2 cyclohexane may exist in either a "boat" or a "chair" form. *However, benzene is planar*, indicating that the geometrical form of the C–C bonds is different than those of aliphatic compounds.

Because of the more even distribution of electrons within the ring and the circular shape of electron orbitals, benzene is frequently (and perhaps more correctly) represented as

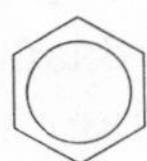

We may use this structure or either of the two resonance forms presented earlier to represent benzene. All are considered correct. However, it must be realized that, electronically, these structures do not represent the whole story.

Benzene is a colorless liquid with an aromatic odor. It is a nonpolar compound, and so it is not soluble in water, but is soluble in hydrophobic organic solvents. Benzene itself is a good organic solvent, burns readily, and is a starting compound for the synthesis of many important chemicals. Unfortunately it is a carcinogen.

10.1 DERIVATIVES OF BENZENE

A *nitro group* (—NO_2) may be added to benzene by heating benzene with nitric and sulfuric acid:

$$C_6H_6 + HNO_3 \xrightarrow{H_2SO_4} C_6H_5NO_2 + H_2O$$

nitrobenzene

Benzene will also react with concentrated sulfuric acid if heated:

$$C_6H_6 + H_2SO_4 \longrightarrow C_6H_5SO_2OH + H_2O$$

benzene sulfonic acid

Halogen atoms may be substituted for hydrogen atoms on the benzene ring:

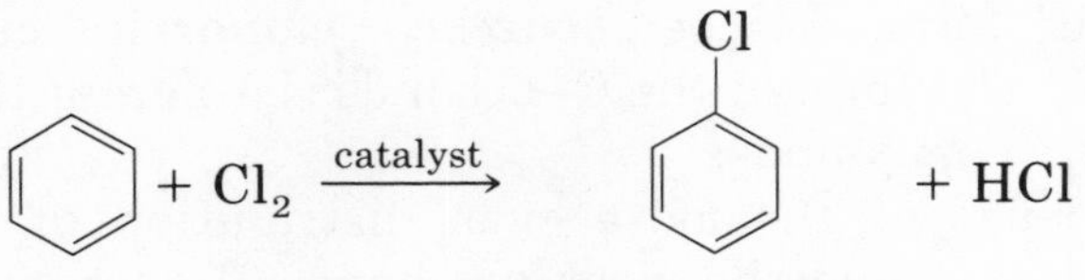

$$C_6H_6 + Cl_2 \xrightarrow{\text{catalyst}} C_6H_5Cl + HCl$$

chlorobenzene

If only one group is attached to the benzene ring the naming of the compounds is easy, but if two substituents are present there are three possible arrangements:

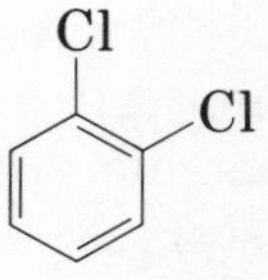

ortho
(*o*-dichlorobenzene)

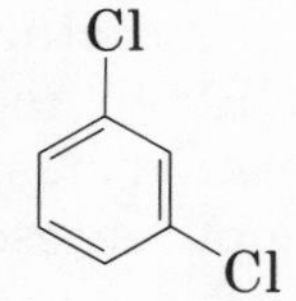

meta
(*m*-dichlorobenzene)

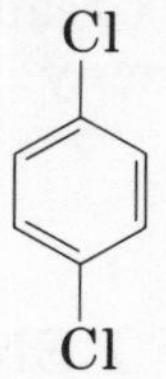

para
(*p*-dichlorobenzene)

By rotating the benzene ring we can see that only these three possible structures are different from one another; hence the following structures are the same as those just given:

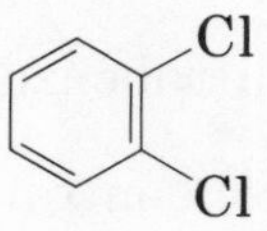

ortho

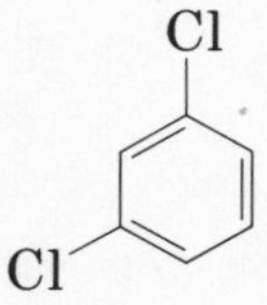

meta

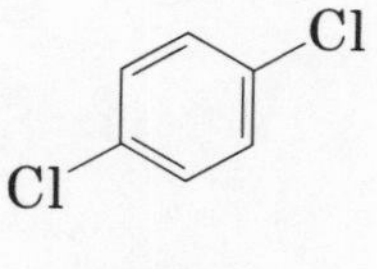

para

A numbering system can be used also, and it is more useful when more than two substituents are present on a ring. Carbon atoms, groups, or radicals attached to the ring are called side chains and they may be numbered in either direction. Again, the smallest numbers possible are used:

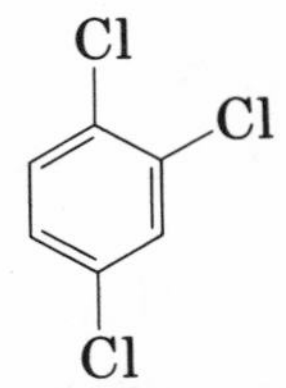

1,2,4-trichlorobenzene 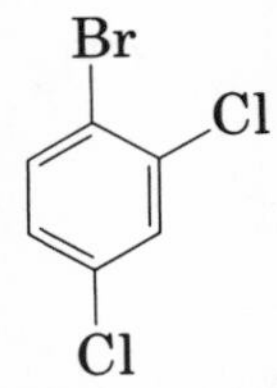 1-bromo-2,4-dichlorobenzene

10.2 ARYL RADICALS

An aryl radical is analogous to an alkyl radical but the former consists of an aromatic ring whereas the latter consists of an aliphatic carbon compound. *Benzene* is a base name but if used as a radical it is called a *phenyl group*.

Toluene is a homolog of benzene and contains an aliphatic radical (methyl) and an aryl radical (phenyl):

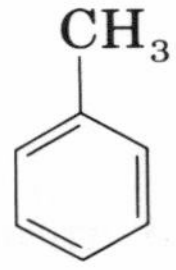

CH_3—

toluene 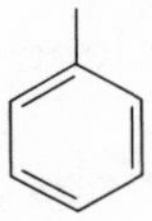 *a phenyl radical* *a methyl radical*

Thus toluene can be called methylbenzene or phenylmethane. Toluene, like benzene, is a good organic solvent and is a useful starting compound of synthetic chemicals. Trinitrotoluene (TNT), or more properly 2,4,6-trinitrotoluene, is a derivative of toluene and quite famous as an explosive:

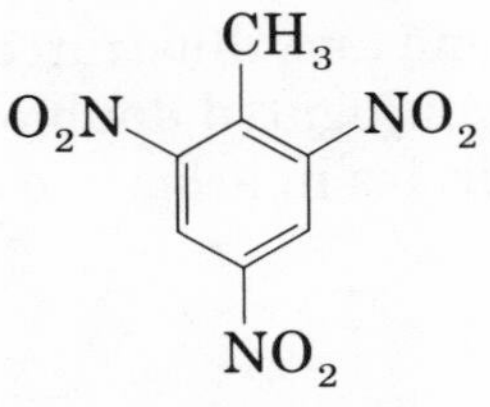

TNT

The substitution of a methyl group for a hydrogen atom on the ring of toluene would yield dimethylbenzene, commonly called xylene. Since the substitution could occur at three different positions on the ring, three isomers of dimethylbenzene exist: 1,2-dimethylbenzene (*o*-xylene), 1,3-dimethylbenzene (*m*-xylene), and 1,4-dimethylbenzene (*p*-xylene):

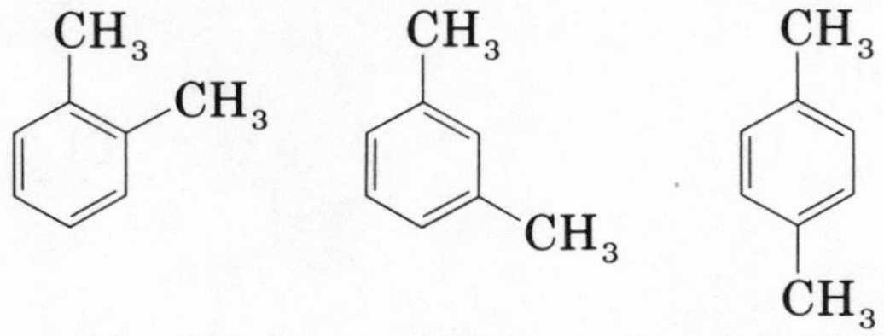

CH_3 CH_3 CH_3 CH_3 CH_3 CH_3

o-xylene *m*-xylene *p*-xylene

The xylenes are useful organic solvents and starting compounds in organic syntheses.

10.3 MULTIPLE RINGS

The simplest condensed (having two or more fused rings) aromatic ring system is naphthalene (mothballs):

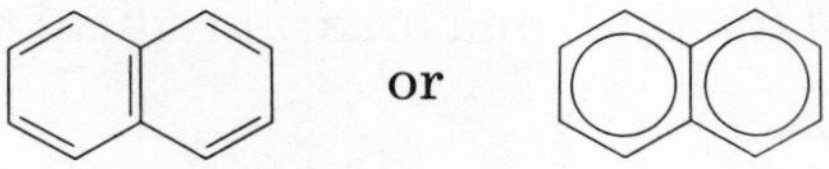

or

naphthalene

There are several condensed rings in the carcinogen 3,4-benzpyrene:

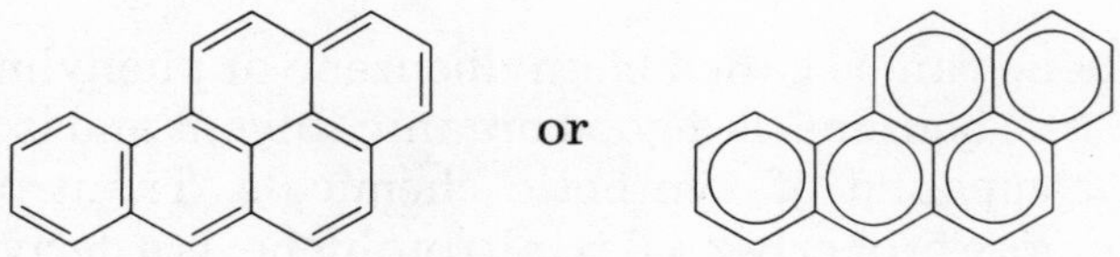

or

3,4-benzpyrene

As described for benzene and illustrated for naphthalene and 3,4-benzpyrene, fully conjugated double-bond systems may be depicted as internal circles in aromatic compounds.

10.4 ARYL ALCOHOLS

Benzyl alcohol may be considered to be derived from toluene. In benzyl alcohol the —OH-group is attached to the aliphatic side chain:

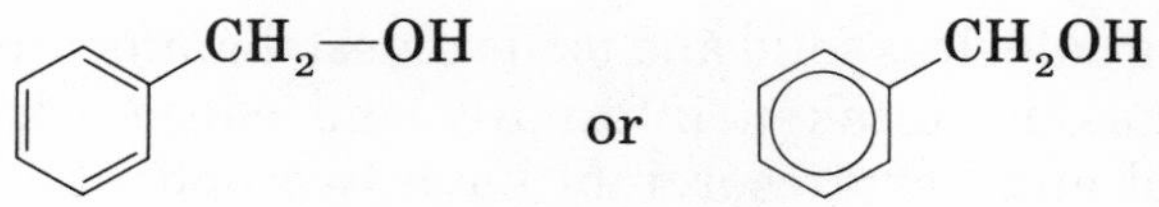

C_6H_5—CH_2—OH or C_6H_5—CH_2OH

benzyl alcohol

Hence it is similar to aliphatic alcohols in many respects. Note that the following group is called a *benzyl group*:

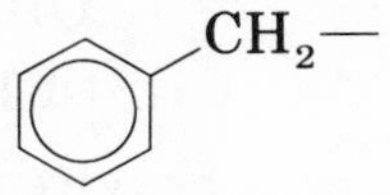

C_6H_5—CH_2—

benzyl group

This is somewhat confusing because of the name benzene, but the benzene ring is called a phenyl group if it is used as a radical:

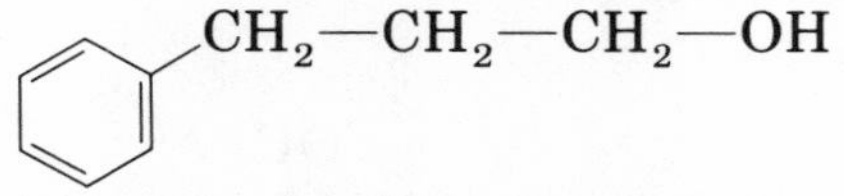

C_6H_5—CH_2—CH_2—CH_2—OH

3-phenyl-1-propanol

10.5 PHENOLS

Although the functional group (—OH) is the same, phenols and aliphatic alcohols have very different properties. A phenol is a compound that has a hydroxyl group substituted for a hydrogen on an *aromatic ring*, whereas aliphatic alcohols have the substitution on an aliphatic chain. The simplest phenol is called phenol, or carbolic acid:

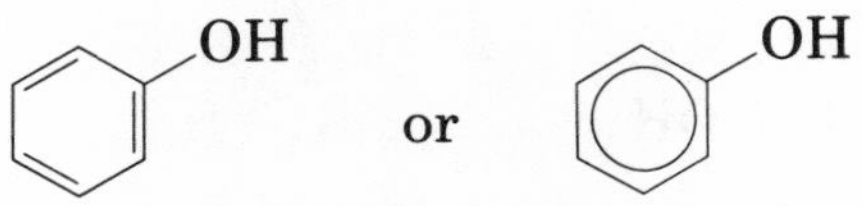

C_6H_5—OH or C_6H_5—OH

phenol

Unlike aliphatic alcohols, phenols are weak acids:

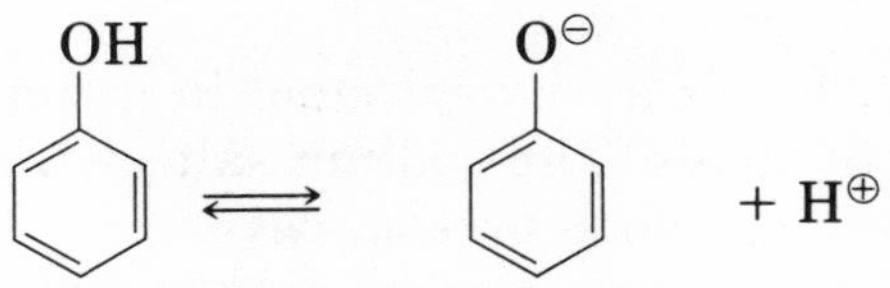

$$C_6H_5OH \rightleftharpoons C_6H_5O^{\ominus} + H^{\oplus}$$

phenol phenolate ion

Phenol is a colorless solid and is the oldest disinfectant known. It is poisonous if taken internally and causes skin burns. Resorcinol and hexylresorcinol have two phenolic hydroxyl groups and are less toxic antiseptics:

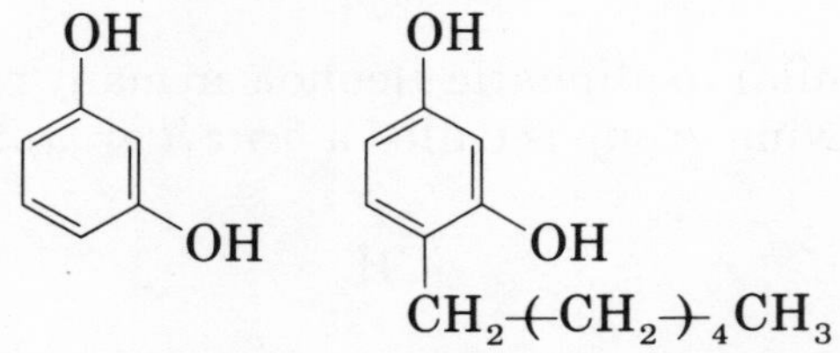

OH OH

OH OH

$CH_2(CH_2)_4CH_3$

resorcinol hexylresorcinol

Cresols are antiseptics and are related to toluene as well as to phenol:

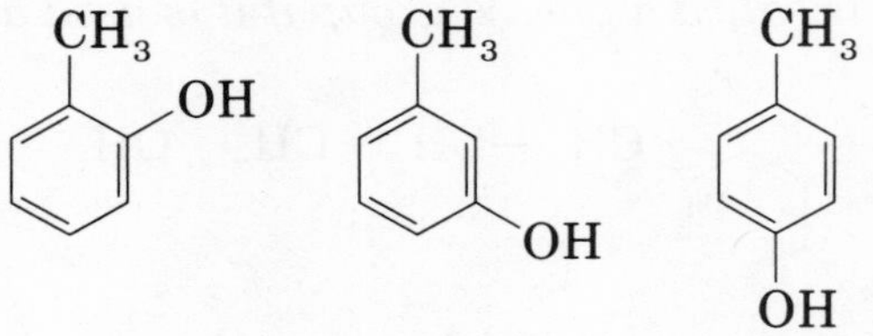

CH_3 CH_3 CH_3

OH

OH

OH

o-cresol *m*-cresol *p*-cresol

Lysol is a soap emulsion of a mixture of these three isomeric cresols.

10.6 ARYL CARBONYL COMPOUNDS

Toluene may be oxidized to form benzoic acid. The conversion and intermediate oxidation states are shown by

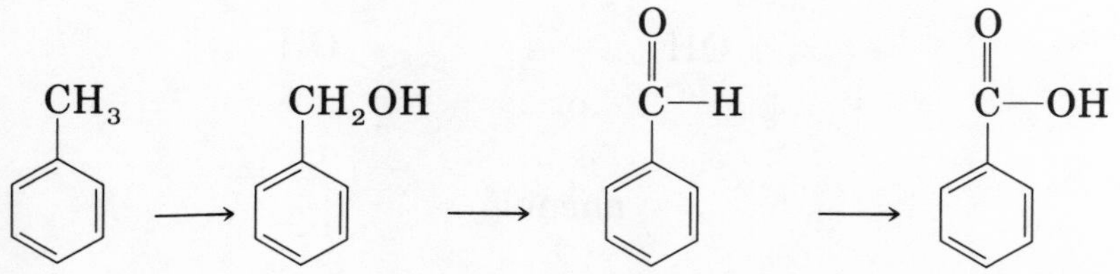

O O

CH_3 CH_2OH C—H C—OH

⟶ ⟶ ⟶

toluene benzyl alcohol benzaldehyde benzoic acid

Benzoic acid behaves as a weak acid in a similar manner as other carboxylic acids. The sodium salt of this acid, sodium benzoate, is used as a food preservative.

Some important aromatic acids include derivatives of salicylic acid, which is *o*-hydroxybenzoic acid:

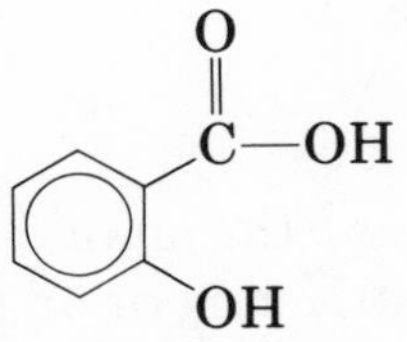

salicylic acid

When the hydroxyl group (or phenol group) is acetylated (when the acetate ester is formed with the —OH-group), the product is aspirin:

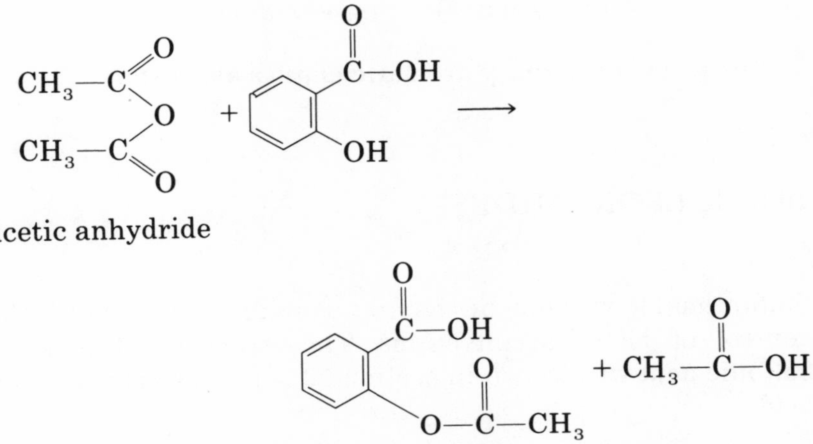

acetic anhydride + salicylic acid ⟶

acetyl salicylic acid
(*aspirin*)

+ CH_3—C(=O)—OH

acetic acid

Methyl salicylate is oil of wintergreen:

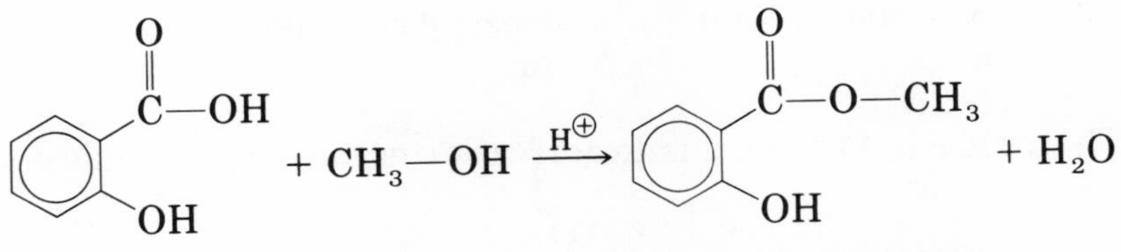

salicylic acid + CH_3—OH $\xrightarrow{H^{\oplus}}$ methyl salicylate + H_2O

methyl salicylate

In addition to forming esters, aromatic carboxylic acids may form amides completely analogous to the aliphatic amides. Yes, primary, secondary, and tertiary amides of aromatic acids exist:

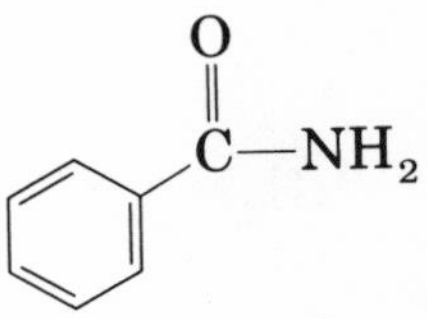

benzamide
(*a primary amide*)

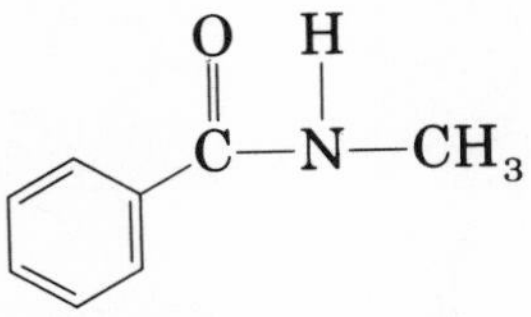

N-methylbenzamide
(*a secondary amide*)

10.7 ARYL AMINES

Like the hydroxyl group, the amino group $—NH_2$ may be substituted on the aromatic ring or on an aliphatic side chain:

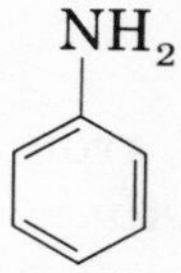

NH_2 $CH_3—CH_2—NH$

aniline
(*a primary amine*)

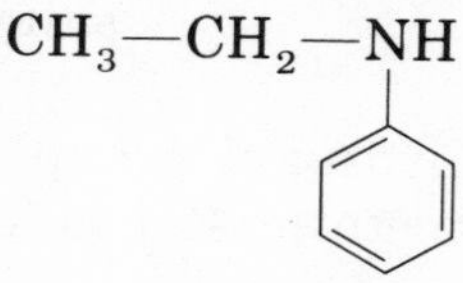

phenylethylamine
(*a secondary amine*)

Tertiary and quaternary aromatic amines also exist.

10.8 SULFONAMIDES

Sulfanilamide was one of the first wonder drugs used for the control of bacterial infections. The structure of benzene sulfonic acid was shown in section 10.1; it does behave as an acid:

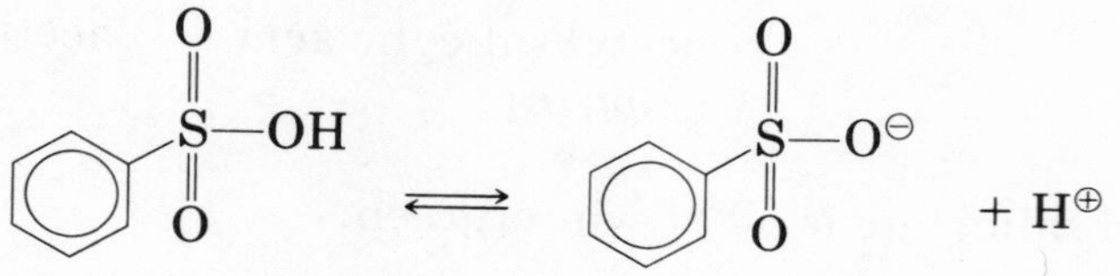

$$C_6H_5—SO_2—OH \rightleftharpoons C_6H_5—SO_2—O^{\ominus} + H^{\oplus}$$

benzene sulfonic acid

benzene sulfonate ion

This acid can form an amide derivative as carboxylic acids do:

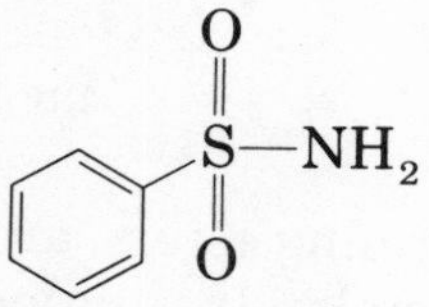

$$C_6H_5—SO_2—NH_2$$

benzene sulfonamide

Sulfanilamide is an aniline and a sulfonamide:

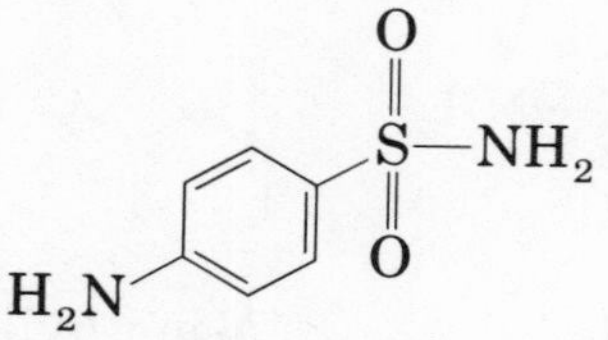

$$H_2N—C_6H_4—SO_2—NH_2$$

sulfanilamide

A family of useful drugs (the sulfas) have been developed where different radicals (R-groups) have been substituted for a hydrogen on the amino or amide groups (secondary amine or amide). They act by blocking the natural action of *p*-aminobenzoic acid (PABA), which many microorganisms need in order to live:

$$\text{R—NH—C}_6\text{H}_4\text{—S(=O)}_2\text{—NH—R} \qquad \text{H}_2\text{N—C}_6\text{H}_4\text{—C(=O)—OH}$$

sulfanilamide derivatives   *p*-aminobenzoic acid

10.9 BIOLOGICAL ROLE OF AROMATIC COMPOUNDS

Aromatic compounds are not nearly as common in biological reactions as aliphatic compounds; in fact the human body has a difficult time handling some of them. However, *some very important compounds* such as *hormones*, a few *amino acids*, and *nucleic acids* do contain aromatic rings:

$$\text{C}_6\text{H}_5\text{—CH}_2\text{—CH(NH}_2\text{)—CO}_2\text{H}$$

phenylalanine, or
3-phenyl-2-aminopropanoic acid
(*an essential amino acid*)

10.10 HETEROCYCLIC COMPOUNDS

Heterocyclic compounds have rings that contain some elements in addition to carbon at one or more positions in the ring. Usually N, O, and S are the elements that replace carbon. Heterocyclic compounds may or may not be aromatic (have conjugated double bonds).

Common oxygen heterocycles include

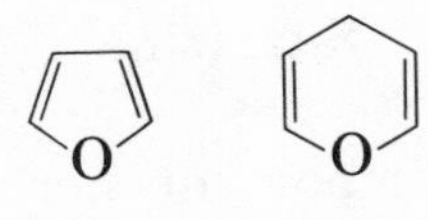

furan   pyran

Note that if an element other than carbon is present, its symbol is always written at the vertex of the polygon where it appears. Hence furan and pyran are five- and six-membered rings that have oxygen replacing carbon in one position only; neither compound is an aromatic compound.

The following are two common nitrogen heterocycles:

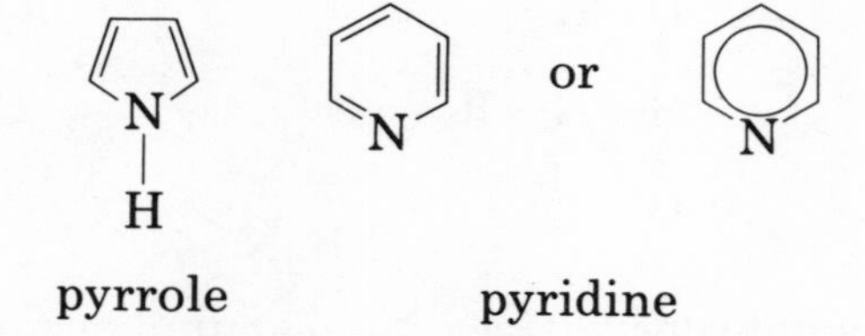

pyrrole pyridine

In naming heterocyclic compounds the hetero-atom (N, O, S, etc.) becomes the #1 position in the ring and groups attached are indicated by numbering in either direction. Of course the lowest numbers possible are used. In open-chain aliphatic compounds hetero-atoms (atoms other than carbon) are not numbered; thus the numbering in heterocyclic compounds is an exception to the usual rule. Note that pyridine is an aromatic compound, but pyrrole is not:

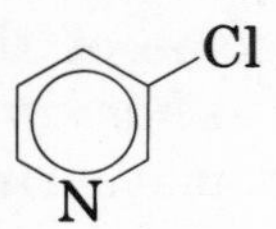

3-chloropyridine, *not* 5-chloropyridine

Several biologically important compounds have two nitrogen atoms in a ring system:

H—N N   N N   or   N N

imidazol pyrimidine

The imidazol ring is present in the amino acid histidine and the aromatic pyrimidine ring system is found in the nucleotide moieties of RNA and DNA.

Three nitrogen atoms, symmetrically arranged (indicated by S) in a six-membered ring, form S-triazine:

N N N   or   N N N

S-triazine

Several commercially important herbicides contain an S-triazine ring.

Two important condensed ring systems are purine and indole:

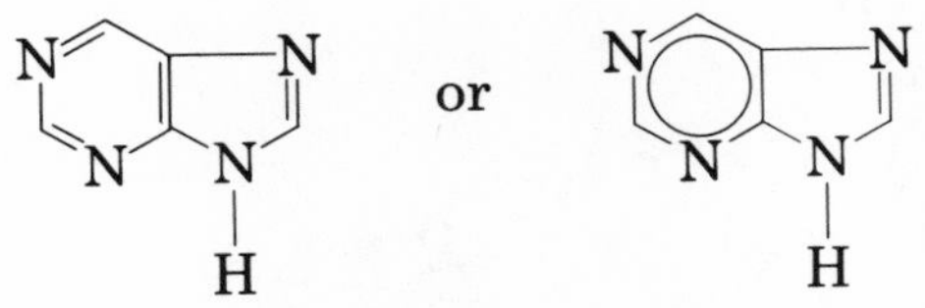

purine

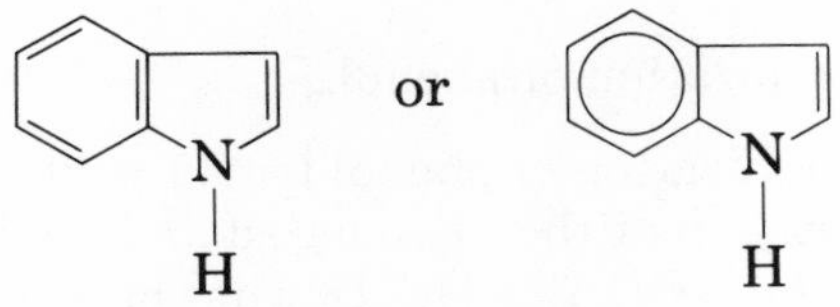

indole

Purine rings are present in nucleotides that make up RNA and DNA, and the indole ring system is a component of the amino acid tryptophan and a plant hormone indolacetic acid (IAA).

Sulfur may also be in a ring, as in the vitamin biotin. In this molecule two 5-member rings are fused, one with two nitrogen atoms and one with a sulfur atom:

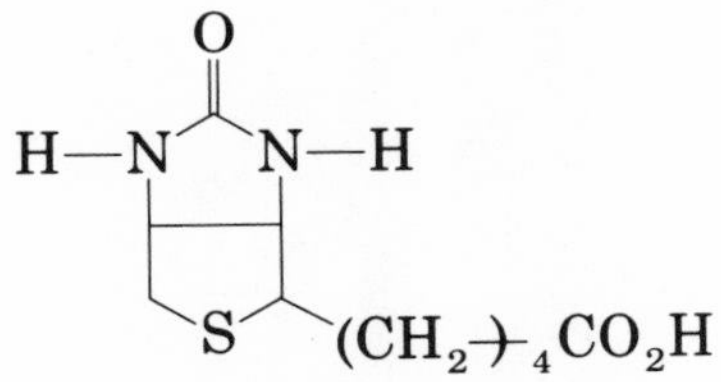

$(CH_2)_4CO_2H$

biotin

EXERCISES

1. Show the structures of the following compounds.
   (a) methyl benzoate
   (b) 3-phenylbutanamide
   (c) *p*-chlorotoluene
   (d) 1,2,4-trichlorobenzene
2. Draw the two resonance forms for the phenyl group of toluene.
3. Name the following compounds.

   (a) 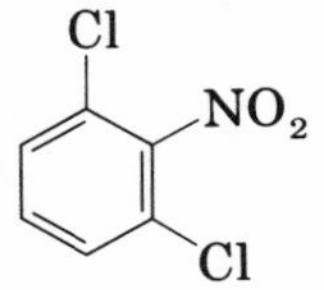 Cl, NO_2, Cl

(b) 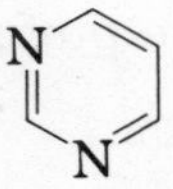 N N

(c) 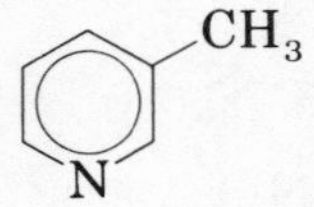 CH_3 N

(d) 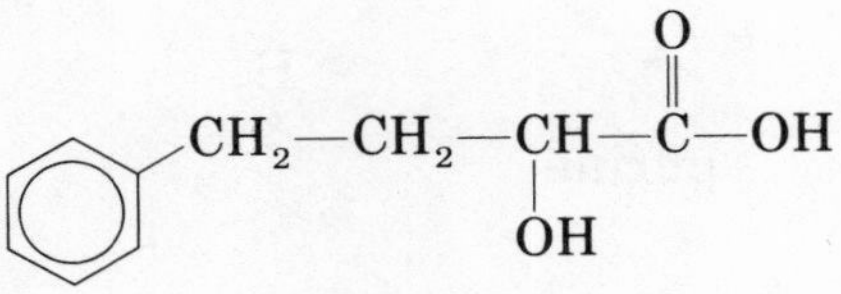 $CH_2—CH_2—CH(OH)—C(=O)—OH$

4. Illustrate the ionization of phenol.

5. The K_a for the ionization of phenol is 1.30×10^{-10}. In a solution, if the concentration of the dissociated and undissociated forms of phenol are equal ($[HA] = [A^{\ominus}]$), what must the pH be?

6. Compare the pK_a for benzoic acid (4.20) to that determined for phenol in question 5. Is benzoic acid a stronger acid than phenol? A strong acid ionizes at a lower pH. Therefore an acid that is 50% ionized at a pH of 3 is much stronger than one that is 50% ionized at a pH of 6. Support your answer with calculations or your reasoning.

Polymers

11.1 SYNTHETIC POLYMERS

Monomer is the general name for a compound that may form complex molecules by the combination of two or more molecules. The combination of *two* monomers (or monomeric units) yields a *di*mer. *Tri*mers have *three* monomeric units and *tetra*mers have four. *Oligo-* means "several"; so *oligomers* consist of several, although an unspecified number of, monomers. *Poly-* means "many," and thus the name *polymer* suggests a large molecule consisting of many smaller units or residues. Copolymers are polymers consisting of two different monomeric units.

To those persons familiar with organic functional groups, the term *polyester* suggests a large molecule containing many ester groups. To those interested in clothing, the word implies a kind of fabric. Both meanings are correct, as will be shown here by the description of the structure of the polyester called Dacron. Dacron is a copolymer of terephthalic acid and ethylene glycol:

$$HO{-}\overset{\overset{O}{\|}}{C}{-}C_6H_4{-}\overset{\overset{O}{\|}}{C}{-}OH \qquad HO{-}CH_2{-}CH_2{-}OH$$

terephthalic acid ethylene glycol

Since acids can form esters with alcohols, one of the carboxyl groups of the acid could react with one of the alcohol groups of ethylene glycol to form an ester linkage between the two molecules. The individual steps involved in making such an ester linkage will not be described; instead net reactions will be illustrated. (You may wish to consult Section A.1 of the Appendix on writing reaction sequences if the system described is not clear.) As shown in Section 5.1 esters can be made by removing a molecule of water from the alcohol and acid. Thus the first linkage can be formed as shown here:

$$HO-\overset{O}{\overset{\|}{C}}-C_6H_4-\overset{O}{\overset{\|}{C}}-O-H + HO-CH_2-CH_2-OH$$

$$\downarrow H_2O$$

$$HO-\overset{O}{\overset{\|}{C}}-C_6H_4-\overset{O}{\overset{\|}{C}}-O-CH_2-CH_2-OH$$

If another molecule of terephthalic acid were available, it could react with the remaining OH-group of the ethylene glycol:

$$HO-\overset{O}{\overset{\|}{C}}-C_6H_4-\overset{O}{\overset{\|}{C}}-O-CH_2-CH_2-OH + H-O-\overset{O}{\overset{\|}{C}}-C_6H_4-\overset{O}{\overset{\|}{C}}-OH$$

$$\downarrow H_2O$$

$$HO-\overset{O}{\overset{\|}{C}}-C_6H_4-\overset{O}{\overset{\|}{C}}-O-CH_2-CH_2-O-\overset{O}{\overset{\|}{C}}-C_6H_4-\overset{O}{\overset{\|}{C}}-OH$$

Next, a second molecule of ethylene glycol could be added:

$$HO-\overset{O}{\overset{\|}{C}}-C_6H_4-\overset{O}{\overset{\|}{C}}-O-CH_2-CH_2-O-\overset{O}{\overset{\|}{C}}-C_6H_4-\overset{O}{\overset{\|}{C}}-O-H + HO-CH_2-CH_2-OH$$

$$\downarrow H_2O$$

$$HO-\overset{O}{\overset{\|}{C}}-C_6H_4-\overset{O}{\overset{\|}{C}}-O-CH_2-CH_2-O-\overset{O}{\overset{\|}{C}}-C_6H_4-\overset{O}{\overset{\|}{C}}-O-CH_2-CH_2-OH$$

By alternately adding terephthalic acid and ethylene glycol molecules, a long chain could be built, ending only when the reactants were exhausted. The esterification reaction can and does take place at both ends of the molecule; therefore the polymeric chain grows from both ends, although the preceding reactions illustrate growth from just one end. Consider the last reaction. The nascent polymer has two free carboxyl groups. To which would ethylene glycol add? As long as ethylene glycol were available we should expect it to be added to both ends. A similar situation exists for the addition of terephthalic acid. The resulting long chains and others like it make up the fibers of Dacron, which eventually are woven into cloth.

Nylon is a polyamide and is another example of a polymeric man-made fiber. The extraordinary variety of plastics available is due to the ingenuity of chemists in making different polymers to fit specific purposes.

11.2 NATURAL POLYMERS

Consider the possibility of forming a polymer from just one kind of molecule such as 2-aminoethanoic acid, or glycine, which is a natural amino acid:

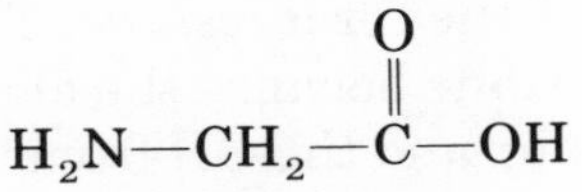

$$H_2N{-}CH_2{-}\overset{\displaystyle O}{\overset{\|}{C}}{-}OH$$

glycine

In the polyester example each monomeric unit had two functional groups, either two carboxyl groups or two alcohol groups. To make a polymer, there *must be at least two* functional groups in each monomer, but they do not have to be the same. Two molecules of glycine can be connected by an amide linkage:

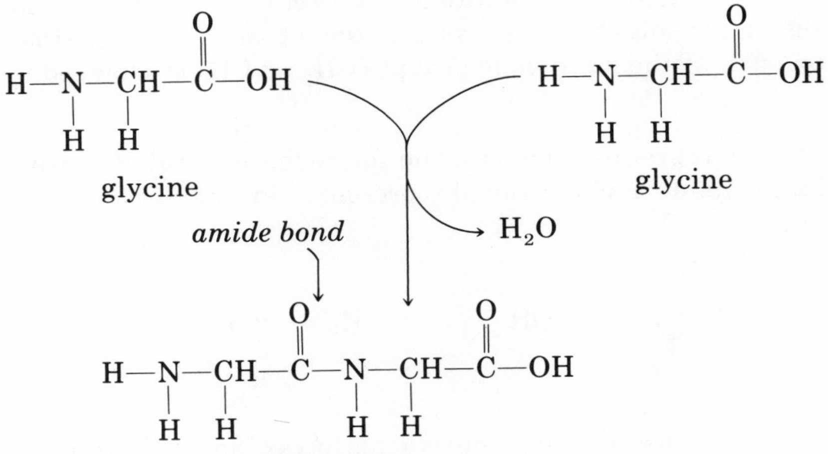

$$\underset{\text{glycine}}{H{-}\underset{\displaystyle H}{\underset{|}{N}}{-}\underset{\displaystyle H}{\underset{|}{CH}}{-}\overset{\displaystyle O}{\overset{\|}{C}}{-}OH} \quad + \quad \underset{\text{glycine}}{H{-}\underset{\displaystyle H}{\underset{|}{N}}{-}\underset{\displaystyle H}{\underset{|}{CH}}{-}\overset{\displaystyle O}{\overset{\|}{C}}{-}OH} \longrightarrow H_2O$$

amide bond

$$\underset{\text{glycylglycine}}{H{-}\underset{\displaystyle H}{\underset{|}{N}}{-}\underset{\displaystyle H}{\underset{|}{CH}}{-}\overset{\displaystyle O}{\overset{\|}{C}}{-}\underset{\displaystyle H}{\underset{|}{N}}{-}\underset{\displaystyle H}{\underset{|}{CH}}{-}\overset{\displaystyle O}{\overset{\|}{C}}{-}OH}$$

The product is a secondary amide and the *secondary amide linkage* established is called a *peptide bond* because such linkages occur in peptides (small proteins) and proteins. By the same net reaction illustrated, glycine units could be added to the dimer, at either or both ends of the molecule. The addition of many glycine units would yield a long chain of glycine residues joined by peptide bonds, which is called polyglycine.

$$\mathrm{H{-}N(H){-}CH(H){-}C({=}O){-}N(H){-}CH(H){-}C({=}O){-}OH} \xrightarrow[-X\,H_2O]{+X\text{ glycine}} \mathrm{H{-}N(H){-}CH(H){-}C({=}O)}\left[\mathrm{{-}N(H){-}CH(H){-}C({=}O){-}}\right]_X\mathrm{N(H){-}CH(H){-}C({=}O){-}OH}$$

polyglycine

Note the use of the term *residue*. The repeating unit in the polymer is not glycine because the elements of water (one H from the amino group and the OH from the acid group) have been eliminated. Thus it is understood that a residue refers to the molecule from which a small inorganic component, such as H_2O, H, or OH, has been removed. The terms *moiety* and *residue* may be used interchangeably.

We might say that polyglycine contains glycine. Technically, this is not correct; however, to a person who understands the general nature of polymers, the statement would not be misinterpreted and, furthermore, the hydrolysis of polyglycine would yield glycine.

The general structure of a protein is the same as that shown for polyglycine *except* that one of the hydrogen atoms on each of the methylene groups (CH_2 or $\mathrm{CH{-}H}$) is replaced by various R-groups. Compare the following general structures for an amino acid to examples presented in Chapter 8.

$$\mathrm{R{-}CH(NH_2){-}C({=}O){-}OH} \quad \text{or} \quad \mathrm{H_2N{-}CH(R){-}C({=}O){-}OH}$$

Because different amino acids have different R-groups, a sequence of four amino acid residues in a protein would appear as follows:

$$-\mathrm{NH}-\underset{\underset{\mathrm{R}}{|}}{\mathrm{CH}}-\overset{\overset{\mathrm{O}}{\|}}{\mathrm{C}}-\mathrm{NH}-\underset{\underset{\mathrm{R'}}{|}}{\mathrm{CH}}-\overset{\overset{\mathrm{O}}{\|}}{\mathrm{C}}-\mathrm{NH}-\underset{\underset{\mathrm{R''}}{|}}{\mathrm{CH}}-\overset{\overset{\mathrm{O}}{\|}}{\mathrm{C}}-\mathrm{NH}-\underset{\underset{\mathrm{R'''}}{|}}{\mathrm{CH}}-\overset{\overset{\mathrm{O}}{\|}}{\mathrm{C}}-$$

a portion of a protein chain

Proteins are called natural polymers because they are made in nature. Other important natural polymers include starch, which is a polymer of glucose. Cellulose, the polymer in cotton fibers, is also a polymer of glucose. The difference between starch (readily used by animals) and cellulose (not digested by monogastric animals) is the way in which the glucose residues are linked together. The genetic information of living organisms is encoded in polymers of deoxyribonucleotides, called *DNA*; and a related polymer of ribonucleotides, *RNA*, is involved in the expression of the genetic information.

11.3 CONCLUSION

Industrial and commercial organic chemistry has not received discussion commensurate with its importance in our daily lives, but exploration of that area of chemistry is beyond the scope of this book. Nevertheless, you may find it interesting to see the complex structures and systematic names of some compounds whose common names you may recognize from ingredient labels or from the news media. Representative insecticides, herbicides, drugs and environmental contaminants are given in the appendix (A.2).

As indicated in the preface, the primary purpose of this book is to present the language of organic chemistry in sufficient detail to permit future discussion of biochemistry. To determine if you have acquired the necessary concepts, a practice exam is provided in the appendix (A.4). If you can answer approximately 80% of the questions without referring to the text, you should be ready to undertake a beginning level course in biochemistry.

EXERCISES

1. Show the general structure of polyalanine. Alanine is 2-aminopropanoic acid.

2. Nylon is a *polyamide* commonly made from adipic acid and 1,6-diaminohexane.

$$\underset{\text{adipic acid}}{HO{-}\overset{\overset{\displaystyle O}{\|}}{C}{-}(CH_2)_4{-}\overset{\overset{\displaystyle O}{\|}}{C}{-}OH} \qquad \underset{\text{1,6-diaminohexane}}{H_2N{-}(CH_2)_6{-}NH_2}$$

It contains secondary amide or peptide linkages. Show the structure of the repeating unit in Nylon.

Appendix

A.1 REACTION SEQUENCES

To help reduce the amount of writing, some conventions are used in diagramming reactions. For example, consider the *hydrolysis* of a *peptide* (secondary amide) bond.

System 1.

$$R-\overset{\overset{\displaystyle O}{\|}}{C}-\overset{\overset{\displaystyle H}{|}}{N}-R' + H_2O \xrightarrow{\text{6N HCl}} R-\overset{\overset{\displaystyle O}{\|}}{C}-OH + H-\overset{\overset{\displaystyle H}{|}}{N}-R'$$

As mentioned before, the reaction is called hydrolysis because water is added to break the bond between the carbonyl carbon atom and the nitrogen atom. In the process, one of the hydrogen atoms of water is separated from the oxygen atom, and then new bonds are formed. The OH from the water is joined to the carbon atom and the H from the water is joined to the nitrogen. The "6N HCl" written above the arrow indicates that the HCl is acting only as a catalyst. The 6N HCl could just as well be written below the arrow.

System 1 is the typical balanced equation shown, and it can become cumbersome when complex molecules and series of reactions are given.

System 2.

$$R-\overset{\overset{O}{\|}}{C}-\overset{\overset{H}{|}}{N}-R' \xrightarrow[+H_2O]{6N\ HCl} R-\overset{\overset{O}{\|}}{C}-OH + H-\overset{\overset{H}{|}}{N}-R'$$

System 3.

$$R-\overset{\overset{O}{\|}}{C}-\overset{\overset{H}{|}}{N}-R' \xrightarrow[H_2O \nearrow]{6N\ HCl} R-\overset{\overset{O}{\|}}{C}-OH + H-\overset{\overset{H}{|}}{N}-R'$$

In both systems 2 and 3 the catalyst is shown and it is indicated that one molecule of water is added during the reaction. In system 2 the "+" in front of the H_2O suggests that the water is added whereas there is nothing about the 6N HCl to suggest that it behaves as a reactant. It is indicated that water is added in system 3 by drawing a line to form a double-ended arrow from the reactants to the products.

System 4.

$$R-\overset{\overset{O}{\|}}{C}-\overset{\overset{H}{|}}{N}-R' \xrightarrow[H_2O]{6N\ HCl} R-\overset{\overset{O}{\|}}{C}-OH \quad + \quad H-\overset{\overset{H}{|}}{N}-R'$$

System 4 is just an extension of system 3. By the use of a common portion of the reaction arrow, two barbed ends and two smooth ends, it is shown that the two reactants combine to yield two products. This arrow system can be used with one, two, three, or more heads and one, two, three, or more tails.

This arrow system is particularly helpful when a sequence of reactions is diagrammed. It permits particular attention to be drawn to certain products or reactants rather than others, and as compared to system 1 the structure of a product does not have to be rewritten when it serves as a reactant in a subsequent reaction:

A → C → D (B enters A → C; E leaves C → D; D and F → G and H; G and I → K and J)

A.2 STRUCTURES OF SOME COMPLEX ORGANIC COMPOUNDS

The names of some complex organic compounds are almost household words. Now that you have some knowledge of organic chemistry, you may wish to see some of the various kinds of structures represented by pesticides, drugs, and environmental contaminants.

A.2.1 *Some Insecticides*

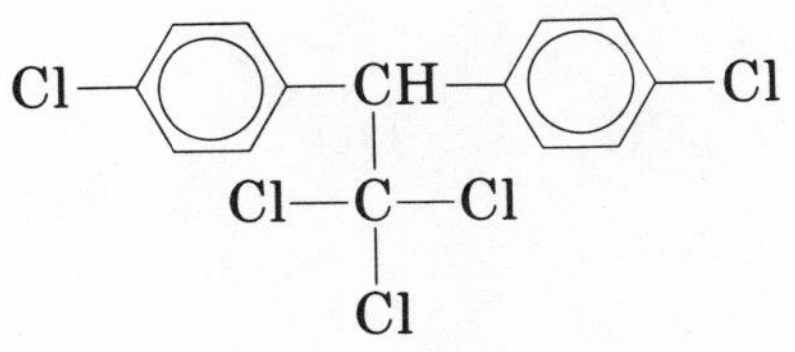

Cl—C₆H₄—CH—C₆H₄—Cl

Cl—C—Cl

Cl

DDT,
or
1,1-bis-(p-chlorophenyl)-2,2,2-trichloroethane

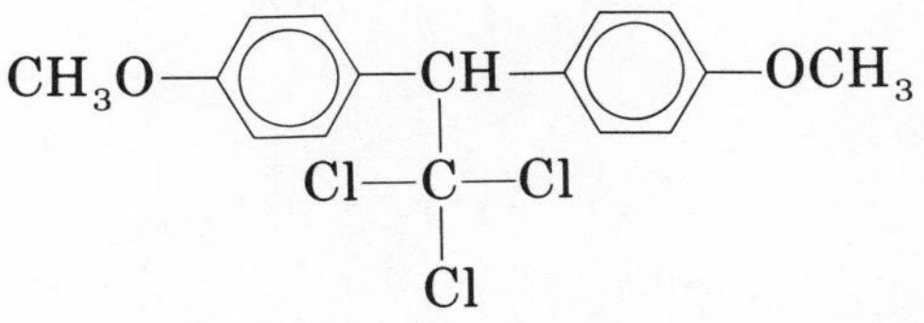

CH_3O—C₆H₄—CH—C₆H₄—OCH_3

Cl—C—Cl

Cl

methoxychlor,
or
1,1′-(2,2,2-trichloroethylidene)bis(4-methoxybenzene)

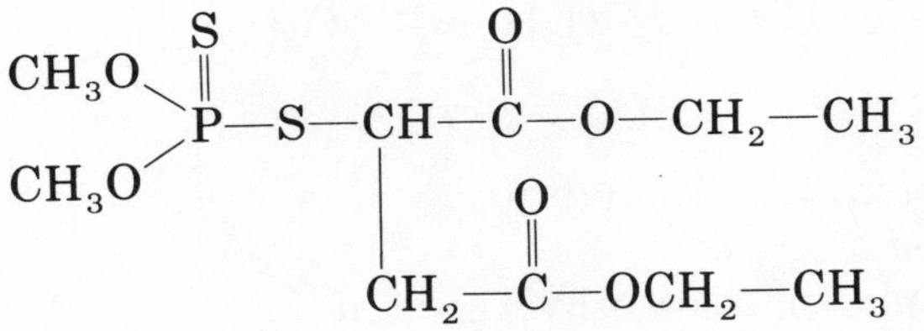

$(CH_3O)_2P(=S)—S—CH(—CH_2—C(=O)—OCH_2—CH_3)—C(=O)—O—CH_2—CH_3$

malathion,
or
S-(1,2-dicarboxyethyl)*o*,*o*-dimethyl-dithiophosphate

A.2.2 *Some Herbicides*

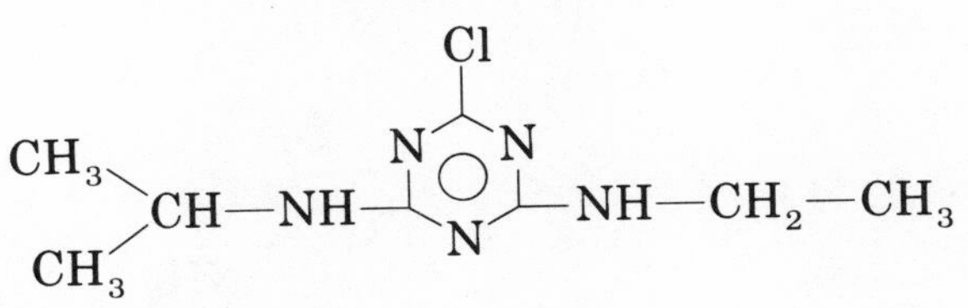

$(CH_3)_2CH—NH—C_3N_3Cl—NH—CH_2—CH_3$

atrazine,
or
2-chloro-4-ethylamino-6-isopropyl-S-triazine

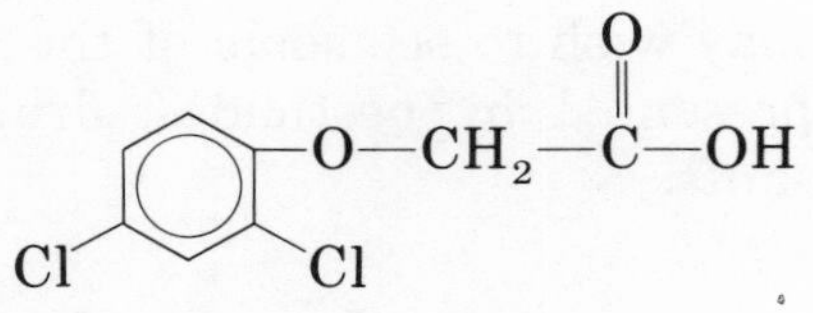

2,4-D,
or
(2-4-dichlorophenoxy) acetic acid

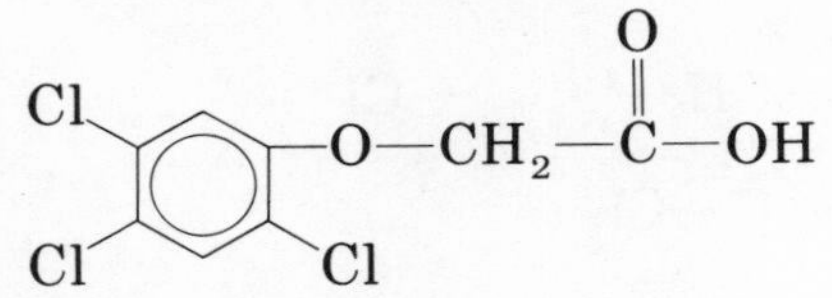

2,4,5-T,
or
(2,4,5-trichlorophenoxy) acetic acid

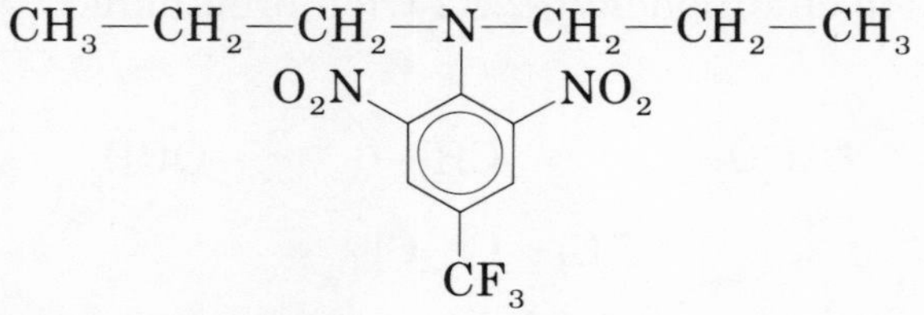

trifluralin (Treflan),
or
α,α,α,trifluoro-2,6-dinitro-*N*,*N*-dipropyl-*p*-toluidine

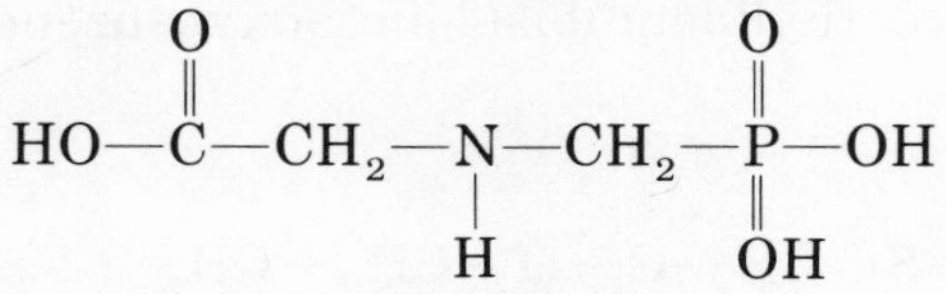

$HO-C(=O)-CH_2-NH-CH_2-P(=O)(OH)-OH$

glyphosate (Roundup),
or
N-(phosphonomethyl) glycine

A.2.3 Some Drugs

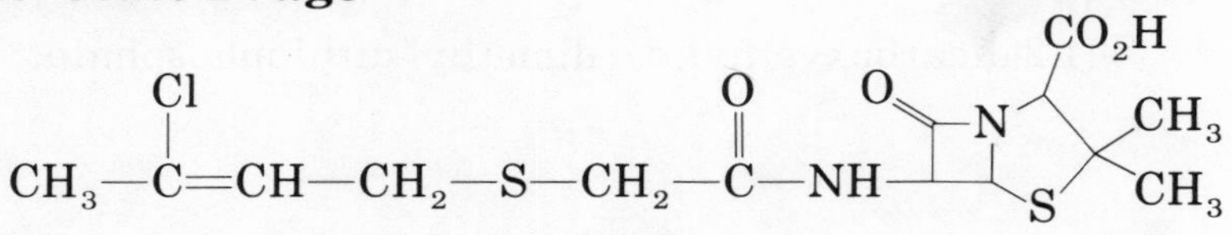

penicillin S

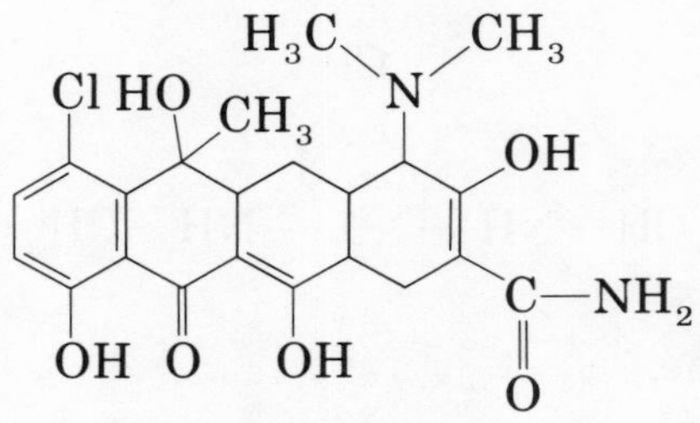

aureomycin

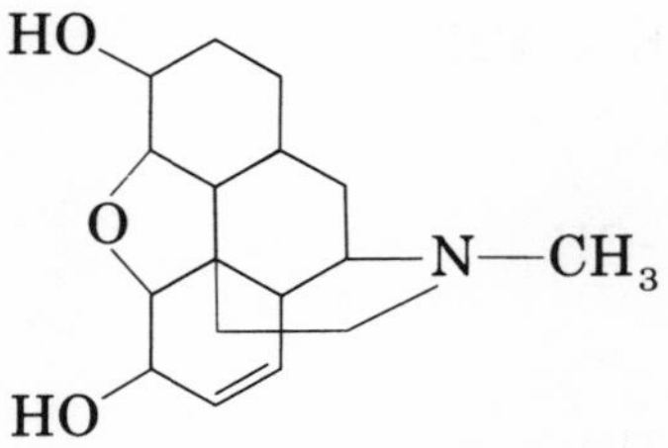

morphine, from opium

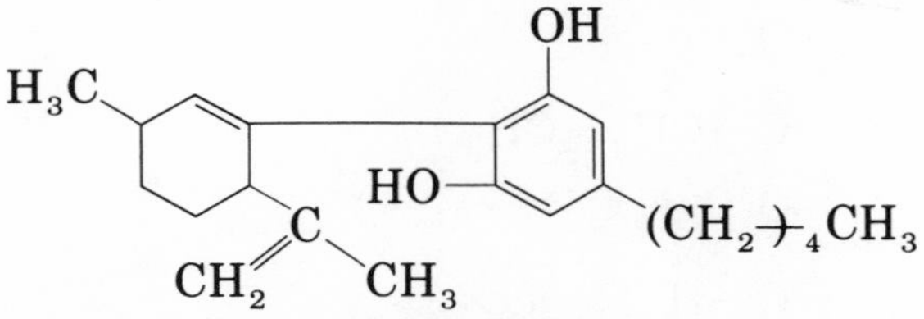

cannabidiol, from marijuana

A.2.4 Some Compounds of Environmental Concern

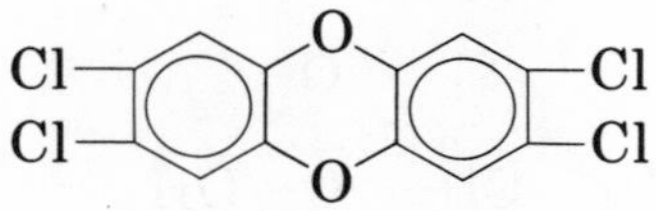

dioxin,
or
2,3,7,8-tetrachlorodibenzo-*p*-dioxin

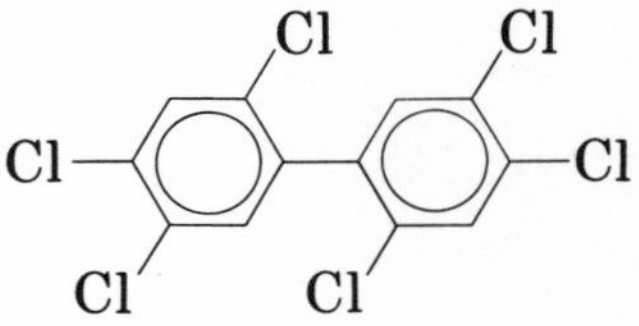

2,4,5,2′,4′,5′-hexachlorobiphenyl

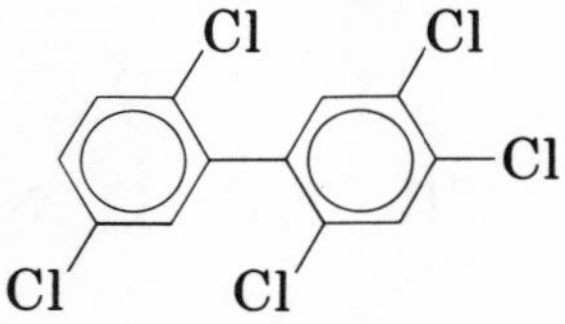

2,4,5,2′,5′-pentachlorobiphenyl

Two examples of PCBs (polychlorinated biphenyls)

A.3 EXTRA PRACTICE NAMING COMPOUNDS

You may wish to sharpen your skills in naming organic compounds. The following exercises are provided for that purpose. Answers are given on pages 143–144.

1. $CH_3{-}CH_2{-}CH_2{-}CH_2{-}CH_3$
2. $CH_3{-}CH_2{-}CH_2{-}\underset{\displaystyle CH_3}{\underset{|}{CH}}{-}CH_3$
3. $CH_2{=}CH{-}CH_2{-}CH{=}CH{-}CH_3$

4. $$CH_2{=}CH{-}\underset{\overset{|}{CH_3}}{CH}{-}CH{=}CH{-}CH_2{-}CH_3$$

5. $$CH_3{-}CH_2{-}\underset{\overset{|}{OH}}{CH}{-}CH_2{-}CH_2{-}CH_3$$

6. $$CH_3{-}CH_2{-}\underset{\overset{|}{OH}}{CH}{-}CH_2{-}CH{=}CH_2$$

7. $$CH_3{-}CH_2{-}CH_2{-}\overset{\overset{O}{\|}}{C}{-}CH_2{-}CH_3$$

8. $$CH_3{-}CH_2{-}CH_2{-}\overset{\overset{O}{\|}}{C}{-}H$$

9. $$CH_3{-}CH_2{-}CH_2{-}\underset{\overset{|}{OH}}{CH}{-}CH_2{-}\overset{\overset{O}{\|}}{C}{-}H$$

10. $$CH_3{-}CH_2{-}CH_2{-}CH_2{-}\overset{\overset{O}{\|}}{C}{-}OH$$

11. $$CH_3{-}CH_2{-}CH_2{-}\underset{\overset{|}{OH}}{CH}{-}\overset{\overset{O}{\|}}{C}{-}OH$$

12. $$CH_3{-}CH_3{-}\underset{\overset{|}{NH_2}}{CH}{-}\overset{\overset{O}{\|}}{C}{-}NH_2$$

13. $$CH_3{-}CH_2{-}CH_2{-}CH_2{-}\overset{\overset{O}{\|}}{C}{-}O{-}CH_2{-}CH_3$$

14. $$CH_3{-}CH_2{-}CH_2{-}\underset{\overset{|}{OH}}{CH}{-}\overset{\overset{O}{\|}}{C}{-}O{-}CH_2{-}CH_3$$

15. $$CH_3{-}CH_2{-}\underset{\overset{|}{OH}}{CH}{-}CH_2{-}\overset{\overset{O}{\|}}{C}{-}\overset{\overset{H}{|}}{N}{-}CH_3$$

16. $$HO{-}\overset{\overset{O}{\|}}{C}{-}CH_2{-}CH_2{-}\underset{\overset{|}{NH_2}}{CH}{-}\overset{\overset{O}{\|}}{C}{-}OH$$

17. $\underset{OH}{\underset{|}{CH_2}}-\underset{NH_2}{\underset{|}{CH}}-\overset{O}{\overset{\|}{C}}-\overset{H}{\overset{|}{N}}-CH_2-CH_3$

18. $CH_3-\overset{CH_3}{\overset{|}{\underset{CH_3}{\underset{|}{C}}}}-\underset{NH_2}{\underset{|}{CH}}-CH_2-\underset{OH}{\underset{|}{CH}}-\underset{OH}{\underset{|}{CH}}-CH_2-\overset{O}{\overset{\|}{C}}-OH$

19. $CH_3-\underset{CH_3}{\underset{|}{CH}}-CH=CH-\underset{OH}{\underset{|}{CH}}-\overset{O}{\overset{\|}{C}}-CH_2-\overset{O}{\overset{\|}{C}}-O-CH_3$

20. $CH_3-\underset{C_6H_5}{\underset{|}{CH}}-CH_2-\underset{NH_2}{\underset{|}{CH}}-\underset{OH}{\underset{|}{CH}}-CH_2-CH_2-\overset{O}{\overset{\|}{C}}-\overset{H}{\overset{|}{N}}-CH_2-CH_3$

21. $CH_3-\underset{C_6H_3(NO_2)Cl}{\underset{|}{CH}}-CH_2-\underset{OH}{\underset{|}{CH}}-\overset{O}{\overset{\|}{C}}-OH$

Note: Sometimes we need to use parentheses to indicate a substituent. Thus the substituted phenyl group would be 4-(3-nitro-4-chlorophenyl).

Answers

1. pentane
2. 2-methylpentane
3. 1,4-hexadiene
4. 3-methyl-1,4-heptadiene
5. 3-hexanol
6. 4-hydroxy-1-hexene (or 3-hydroxy-5-hexene—higher numbers but acceptable)
7. *n*-propyl ethyl ketone or 3-hexanone
8. butanal or butyraldehyde
9. 3-hydroxyhexanal or β-hydroxycaproaldehyde
10. pentanoic acid or valeric acid
11. 2-hydroxypentanoic acid or α-hydroxyvaleric acid
12. 2-aminobutanamide or α-aminobutyramide
13. ethyl pentanoate or ethyl valerate

14. ethyl 2-hydroxypentanoate or ethyl α-hydroxyvalerate
15. *N*-methyl-3-hydroxypentanamide or *N*-methyl-β-hydroxyvaleramide
16. 2-aminopentanedioic acid, α-aminoglutaric acid, or glutamic acid
17. *N*-ethyl-2-amino-3-hydroxypropanamide or *N*-ethyl-α-amino-β-hydroxypropionamide
18. 6-amino-3,4-dihydroxy-7,7-dimethyloctanoic acid
19. methyl 4-hydroxy-7-methyl-3-oxo-5-octenoate
20. *N*-ethyl-5-amino-4-hydroxy-7-phenyloctanamide
21. 2-hydroxy-4-(3-nitro-4-chlorophenyl)-pentanoic acid

Note: In answers 17 through 20 the alphabetical rules of order for the groups were followed, but the groups could have been listed in order of group complexity. Actually, about any order would be acceptable to most people.

Practice Exam

The following practice exam is offered so that you may determine how well you have mastered the material presented in the preceding chapters.

I. Draw the complete Lewis structure for ethanol, or ethyl alcohol.

II. Using structures, illustrate resonance for the acetate ion.

III. Show the possible structural isomers for propyl chloride (C_3H_7Cl).

IV. Using the correct structures for all compounds that you show, illustrate the reaction called saponification.

V. Name the following compounds.

1. $CH_3{-}CH_2{-}CH_2{-}\overset{\overset{\displaystyle O}{\|}}{C}{-}O{-}CH_2{-}CH_3$ ____________________

2. (benzene ring with CH_3 and Cl on adjacent carbons) ____________________

3. $CH_3{-}CH{=}CH{-}CH_2{-}CH{=}CH_2$ ____________________

4. $CH_3{-}CH_2{-}\underset{\underset{\displaystyle OH}{|}}{CH}{-}CH_2{-}CH_2{-}\overset{\overset{\displaystyle O}{\|}}{C}{-}\overset{\overset{\displaystyle H}{|}}{N}{-}H$ ____________________

VI. Matching. There may be one or several correct answers. Only one correct answer is requested. Not all of the structures will be used and some may be used more than once.

a. 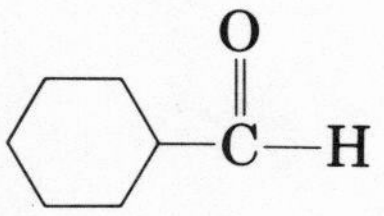 (cyclohexane ring)$-\overset{O}{\overset{\|}{C}}-H$

b. $CH_3-\overset{O}{\overset{\|}{C}}-CH_3$

c. $CH_3-CH_2-NH_2$

d. $CH_3-CH_2-CH_2-CH_2-SH$

e. $CH_3-CH_2-\overset{O}{\overset{\|}{C}}-OH$

f. 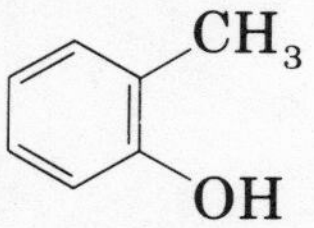 (benzene ring with CH_3 and OH substituents)

g. $CH_3-O-O-CH_3$

h. $CH_3-\overset{\oplus}{N}(CH_3)_2-CH_2-CH_3$

i. $CH_3-\overset{O}{\overset{\|}{C}}-OCH_3$

j. 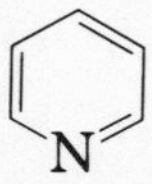 (pyridine ring, N)

k. $CH_3-\underset{OH}{\underset{|}{CH}}-CH_2-CH_3$

l. $CH_3-\underset{OH}{\underset{|}{\overset{CH_3}{\overset{|}{C}}}}-CH_2-CH_3$

m. $CH_3-\overset{O}{\overset{\|}{C}}-\overset{H}{\overset{|}{N}}-CH_3$

n. $CH_3-\overset{O}{\overset{\|}{C}}-NH_2$

o. $CH_3-S-CH_2-CH_3$

p. $CH_3-\overset{CH_3}{\overset{|}{CH}}-CH_3$

1. ___ a ketone
2. ___ a secondary aliphatic alcohol
3. ___ a carboxylic acid
4. ___ a phenol
5. ___ a compound that contains a phenyl radical or group
6. ___ an aldehyde
7. ___ an aromatic compound
8. ___ a mercaptan or thiol
9. ___ a compound that contains a carbonyl group but is not a carboxylic acid
10. ___ a primary amine
11. ___ a compound that contains a quaternary ammonium group
12. ___ an ester
13. ___ a secondary amide
14. ___ a tertiary alcohol
15. ___ a heterocyclic compound

VII. Write the structures for the following compounds.
1. valeric acid
2. malonic acid
3. 3-phenyl-2-aminopropanoic acid
4. 2,6-diaminohexanoic acid

VIII. Write the balanced reaction and complete structures for the hydrolysis of acetylphosphate.

IX. The pK_a for benzoic acid is 4.20. Calculate the pH of a solution of benzoic acid in which the concentration of the undissociated form (HA) is 0.002 M and the concentration of the dissociated form ($A^{\ominus}$) is 0.05 M.

X. Add the appropriate letter or word to indicate which isomer is represented by each of the following structures.

1. 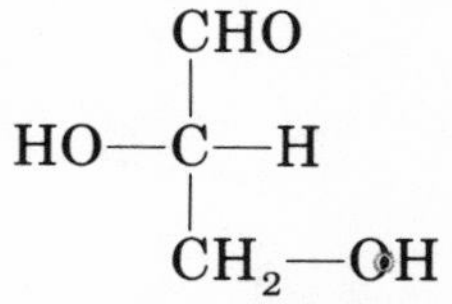
```
        CHO
         |
   HO—C—H
         |
        CH2—OH
```
glyceraldehyde

2. 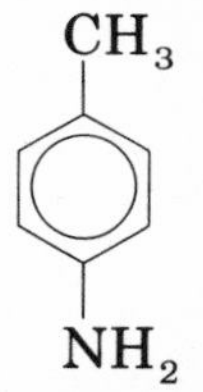
```
      CH3
       |
   (benzene ring)
       |
      NH2
```
aminotoluene

3. 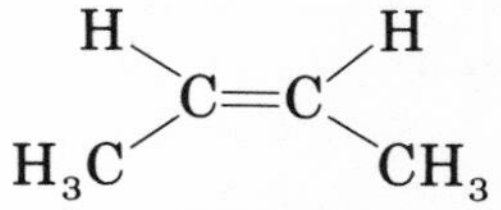
```
   H         H
    \       /
     C = C
    /       \
 H3C         CH3
```
2-butene

Index